In *Physiognomy and the Meaning of Expression in Nineteenth-Century Culture*, Lucy Hartley examines the emergence of physiognomy as a form of popular science. Physiognomy posited an understanding of the inner meaning of human character from observations of physical appearances, usually facial expressions. Taking the physiognomical teachings of Johann Caspar Lavater as a starting-point, Hartley considers the extent to which attempts to read the mind and judge the character through expression can provide descriptions of human nature. She argues that the writings of Charles Bell, and the Pre-Raphaelites establish the significance of the physiognomical tradition for the study of expression whilst preparing the ground for the rise of new doctrines for the expression of emotion by Alexander Bain and Herbert Spencer. She then demonstrates how the evolutionary explanation of expression proposed by Spencer and Charles Darwin is both the outcome of the physiognomical tradition and the reason for its dissolution.

LUCY HARTLEY is Lecturer in English at the University of Southampton. Her articles have appeared in *Journal of Victorian Culture*, *Textual Practice* and *Nineteenth-Century Literature*.

CAMBRIDGE STUDIES IN NINETEENTH-CENTURY
LITERATURE AND CULTURE 29

PHYSIOGNOMY AND THE
MEANING OF EXPRESSION IN
NINETEENTH-CENTURY CULTURE

CAMBRIDGE STUDIES IN NINETEENTH-CENTURY
LITERATURE AND CULTURE 29

General editor
Gillian Beer, *University of Cambridge*

Editorial board
Isobel Armstrong, *Birkbeck College, London*
Leonore Davidoff, *University of Essex*
Terry Eagleton, *University of Oxford*
Catherine Gallagher, *University of California, Berkeley*
D. A. Miller, *Columbia University*
J. Hillis Miller, *University of California, Irvine*
Mary Poovey, *New York University*
Elaine Showalter, *Princeton University*

Nineteenth-century British literature and culture have been rich fields for interdisciplinary studies. Since the turn of the twentieth century, scholars and critics have tracked the intersections and tensions between Victorian literature and the visual arts, politics, social organization, economic life, technical innovations, scientific thought – in short, culture in its broadest sense. In recent years, theoretical challenges and historiographical shifts have unsettled the assumptions of previous scholarly synthesis and called into question the terms of older debates. Whereas the tendency in much past literary critical interpretation was to use the metaphor of culture as 'background', feminist, Foucauldian, and other analyses have employed more dynamic models that raise questions of power and of circulation. Such developments have reanimated the field.

This series aims to accommodate and promote the most interesting work being undertaken on the frontiers of the field of nineteenth-century literary studies: work which intersects fruitfully with other fields of study such as history, or literary theory, or the history of science. Comparative as well as interdisciplinary approaches are welcomed.

A complete list of titles published will be found at the end of the book.

PHYSIOGNOMY AND THE MEANING OF EXPRESSION IN NINETEENTH-CENTURY CULTURE

LUCY HARTLEY

CAMBRIDGE
UNIVERSITY PRESS

PUBLISHED BY THE PRESS SYNDICATE OF THE UNIVERSITY OF CAMBRIDGE
The Pitt Building, Trumpington Street, Cambridge, United Kingdom

CAMBRIDGE UNIVERSITY PRESS
The Edinburgh Building, Cambridge CB2 2RU, UK   www.cup.cam.ac.uk
40 West 20th Street, New York, NY 10011–4211, USA   www.cup.org
10 Stamford Road, Oakleigh, Melbourne 3166, Australia
Ruiz de Alarcón 13, 28014 Madrid, Spain

First published 2001

Printed in the United Kingdom at the University Press, Cambridge

Typeset in Baskerville 11/12.5pt   System 3b2   [CE]

A catalogue record for this book is available from the British Library

ISBN 0 521 79272 x hardback

To my parents

Contents

Plates

The author and publisher are grateful to the institutions above who have kindly granted permission to reproduce pictures.

Acknowledgements

It is a pleasure to record a number of debts. Amongst those who have advised and helped me during my research, I would like to thank Joseph Bristow and Ludmilla Jordanova, supervisors of the doctoral thesis in which some of these ideas first appeared: they provided invaluable guidance and insight as well as encouragement (and sometimes surprise) for what I was trying to do. At various stages of this work, a number of people have given up time to read early drafts, listen to my ramblings, or offer encouragement; for this I would like to thank Stephen Bann, John Barrell, Stephen Bygrave, Caroline Blinder, Cora Kaplan, Peter Middleton and James Moore. My colleagues in the Department of English at Southampton University have also been unfailingly helpful and supportive over the last few years.

I am indebted to the anonymous readers for Cambridge University Press for showing me what this book could be when it was not clear to me. Thanks are due also to the series editor, Gillian Beer, and Linda Bree for their valuable help. I am grateful to the Guildhall Art Gallery (London), the Walker Art Gallery (Liverpool), and the Wellcome Institute Library (London) for granting me permission to reproduce a number of the illustrations in the book.

My greatest debt is to my family, and especially my parents, Winifred and Keith, as they have borne the true cost of the book, not only financially in the early days, but with limitless understanding, enthusiasm, and support – it is to them that I dedicate this book.

Introduction

'Physiognomy, whether understood in its most extensive or confined signification, is the origin of all human decisions, efforts, actions, expectations, fears, and hopes', Johann Caspar Lavater wrote in *Essays on Physiognomy* (1789–93); 'from the cradle to the grave . . . from the worm we tread on to the most sublime of philosophers . . . physiognomy is the origin of all that we do and suffer'.[1] To modern readers, Lavater's claim for physiognomy as the most fundamental form of action seems hopelessly ambitious and perhaps more than faintly ridiculous. Yet it was due to his work that physiognomy was popular throughout the nineteenth century as a means of describing character through expression. Physiognomy was, to Lavater, the root of human actions, sensations, and beliefs because it described and explained the most natural responses of individuals to each other – acts of judgement – and placed them within a religious framework. Suppose, he said, we take the example of a man in the company of a stranger, the man will 'observe, estimate, compare, and judge him, according to appearances, although he might never have heard of the word or thing called physiognomy; [there is] not a man who does not judge of all things which pass through his hands, by their physiognomy; that is, of their internal worth by their external appearance'.[2] The idea was that physiognomy offered a spiritual guarantee that anyone could read the appearances of things in the world and then form a judgement on the basis of their essential though hidden value. If we conceive human nature primarily in terms of self, feelings, and identity, Lavater seems to imply, then our relations with others are such that we instinctively make quite profound judgements of what we see without considering the reasons for doing so. Thus, physiognomy defines and attempts to explain the scope of these instinctive responses, focussing not on why we make such judgements but on the fact that we do, and so places the burden of judgement on the observation of actualities rather than the explanation of causes, with the latter naturally only explicable with reference to a divinity.

A contemporary definition from the fourth edition of *The Encyclopaedia Britannica* (1810) spells this out. Physiognomy, it says, 'is a word formed from the Greek for nature, and I know', and it means 'the knowledge of the internal properties of any corporeal existence from the external appearances. [Physiognomists] among physicians, denote such signs as, being taken from the countenance, serve to indicate the state, disposition, &c. both of the body and mind: and hence the art of reducing these signs to practice is termed physiognomy'.[3] The practice of physiognomy, as defined above, is concerned with natural knowledge; that is to say knowledge which is instinctive and, as such, distinct from that which is learnt or acquired, or, to put it in other words, the product of an involuntary as opposed to a voluntary response. So, the natural in this knowledge seems to imply the kind of knowledge which develops from an instinctive grasp of the correspondences between the external form and the internal properties of living forms and, in particular, human beings. This knowledge is complex but it is, nonetheless, accessible to everyone because it involves what actually exists in the organic world, and requires only that the process of reduction – the seeing of the external as a sign or index of the internal – be learned. On the basis of this definition, physiognomy seems to serve a social as well as a religious function as it posits an understanding of the inner meaning of human nature from the observations of actual appearances – facial expressions are used in this context to tell us about the kind of person we see before us by virtue of the fact that the expression of a specific kind of emotion stands for a standard type of character. That which we take to be peculiar and distinctive to an individual then becomes that which is common and ordinary to all individuals. Conceived in this light, the crux of physiognomic practice is a classificatory act which functions in a profoundly normative manner in so far as it takes a particular expression as the exemplification of a general kind and then uses this to describe the character of an individual.

The fit between particular and general is important because it indicates the strength of physiognomic practice and points also to its weakness. Lavater's goal was to construct a scheme of classification adequate to describing the variety of human nature. Although the aim was to provide an economy of explanation for a large number of natural phenomena, the emphasis of Lavater's work was on the social relations of individuals which were, he claimed, explicable mainly through facial expression. In order to describe individual expressions in terms of general types, he translated the bewildering complexity of particular expressions into simple facial signs and so offered a normative scale of seeing – one which was based, however, on subjective measures. The interest in physiognomical teachings depended on such subjective measures, and Lavater was not alone in proposing explanations of the emotions from the correspondences between expression and emotion.

This book explores changing understandings of expression, primarily the expression of the emotions, and principally via the face, from the English publication of Lavater's *Essays on Physiognomy* (1789) to the publication of Francis Galton's *Hereditary Genius* (1892).[4] What is involved in expression is a complex issue, far removed from what physiognomy in the form developed by Lavater is prepared to acknowledge. The key question is whether we can take a snapshot of an individual based on their appearance or how they look. Physiognomic practice assumes the picture it draws of an individual is truthful because it is derived from what the external appearance of an individual tells us about their internal nature, even though the latter is conceptualised within strict codes. This attempt to describe the core of our nature, or what defines us as human beings, is an age-old pursuit which is at once interesting and perplexing, and continues to arouse an equal amount of ambitious claims and unbridled curiosity in our present age.[5] In its physiognomic form, this pursuit is ultimately doomed to fail as it offers false claims about human existence based on the idea that the character and behaviour of an individual, and his/her activity in society, are explicable through facial expression. Hence, the significance of physiognomy lies in its consequences rather than the reality it constructs. It is through the primitive perspective on human nature which it advances, I suggest, that we can see the tentative beginnings of a tradition of modern psychological thought. What emerges in the later nineteenth and early twentieth century as a psychological account of human character and behaviour – a science of the mind – is both the long-term outcome of physiognomical teachings and the reason for their dissolution. As John Stuart Mill declared in an essay on Alexander Bain's *Senses and*

Intellect (1855), 'The sceptre of psychology has decidedly returned to this island and the scientific study of mind, which for two generations, in many other respects distinguished for intellectual activity, had, while brilliantly cultivated elsewhere, been neglected by our countrymen, is now nowhere prosecuted with so much vigour and success as in Great Britain.'[6] Placed in the context of theories of expression presented by Charles Bell, Alexander Bain, Charles Darwin, and Francis Galton, Lavater's physiognomical teachings assume an important role not only in laying the groundwork for the development of theories of expression derived from physiology, but also in representing the tensions and contradictions of nineteenth-century scientific thought. The marked renewal of interest in physiognomy in the nineteenth century is prompted by the ease which with Lavater's teachings could, with a little help, be learnt from everyday life and then applied to make sense of that life; and, as I shall demonstrate, the fact that Lavater proclaimed this to be a scientific mode of analysis served only to increase its popular appeal.

Physiognomy has a long and chequered history, from the classical tradition of Aristotle, to Giovanni Battista della Porta and Charles Le Brun in the sixteenth and seventeenth centuries, and Lavater, Bell, and Darwin in the nineteenth century.[7] Throughout this time, it seems to have aroused such conflicting emotions that it has been both lauded as a source of knowledge about nature and man, and disparaged as a mystical and highly deterministic practice. However, it is the emergence of physiognomy in the nineteenth century which has received the most critical attention. John Graham's studies of the 1960s serve as invaluable reference works for scholars interested in physiognomy, as they consider the development of Lavater's ideas and trace the importance of physiognomic practice in England in the nineteenth century.[8] Graeme Tytler's later work in the 1980s is also significant as it was probably the first study to draw a clear parallel between physiognomy and literature; the revival of interest in physiognomy after Lavater is the result of 'various cultural forces' which have, Tytler claims, a clear literary bent.[9] Yet, in general, studies of physiognomy have tended to emphasise the use of physiognomy as a hermeneutic praxis over its theoretical foundations, and usually in isolation from contemporary debates on man, mind, and nature.[10] These studies are not unimportant (nor indeed uninteresting) but they only give part of the picture necessary for understanding physiognomy as the impetus for the development of theories of expression.

The nineteenth century was a period in which man's place in nature was the subject of intense and often heated debate, specifically amongst the scientific community.[11] Adrian Desmond's intelligent and insightful study of the debates about the development of the organic world (before Darwin) presents an incredibly detailed picture of the issues, arguments, and figures which enter the arena at various stages of the discussions. It is necessary to have 'an unashamedly political perspective', Desmond argues, to 'examine the reasons why the radicals exploited the doctrines of nature's self-development and how these ideas served their democratic ends'.[12] The richness and density of debate was such that an understanding of the laws in nature did not preclude a belief in the actions of a higher law-maker, though admittedly this was a truly contentious issue. New ideas about the development of life on earth and new explanations of its natural phenomena offered compelling models of the history, structure, and function of the organic world – based for instance on the geology of Charles Lyell, the comparative anatomy of Georges Cuvier, the physiology of W. B. Carpenter and Alexander Bain, or the evolutionary theory of Charles Darwin and Alfred Russel Wallace – and demonstrated that it was no longer necessary to place man at the centre of explanations of change and transformation.[13] But this was also a period which inherited and went on to explain physiognomical teachings which seemed to affirm the purpose and design of a natural order of things. 'The encoding of human types through physiognomy, in art as in life', Mary Cowling claims, 'was a means of bringing order into an ever-increasing, even bewildering variety of human types and social classes: a localized variation of what was being performed on a global scale by anthropologists'.[14] Cowling's study presents physiognomy as a form of hermeneutics which was used frequently in artistic and literary contexts as an important means of characterising subjects. Cowling argues that the popularity of physiognomy derived from its capacity to employ typological forms of classification, and as a result it became an important resource for, and was often appropriated by, mid-nineteenth-century genre painters like William Powell Frith, William Maw Egley, and George Elgar Hicks. Persuasive though Cowling's work is about the use of physiognomy to underwrite literary and artistic practice, the extent to which physiognomy fitted into the wider debates of the time about the organic world is assumed rather than explained. What did the natural order look like? Were there laws in nature? How did these affect (human) action and expression? It is precisely these questions which I shall undertake to explore, with the intention of evaluating physiognomy in the light of new scientific debates about man, mind, and nature. It is not insignificant that physiognomy became popular during the nineteenth century at a time when fundamental questions were being asked of natural phenomena, but, unlike Cowling, I argue that there is a direct relationship between physiognomy and new debates about the structure and function of the world; in my opinion, physiognomy both responds to and reacts against different conceptions of change in the organic world.

There is no doubt that the emergence of physiognomy in the nineteenth century as a popular phenomenon was due in large part to the moral framework it provided for everyday life with all its 'decisions, efforts, actions, expectations, fears, and hopes'. Despite, or more probably because of its claims to moral authority, physiognomy seems to have aroused considerable attention amongst the philosophical and scientific communities of the time, raising a number of issues about the meaning of expression. What is the purpose of expressions? What is the relationship between expressions and emotions? Is an understanding of expressions innate or learned? Does an explanation of the expression of the emotions tell us anything about human nature, character, and behaviour? These are the sort of questions a number of leading thinkers (such as Bain, Darwin, and Galton) asked as they attempted to discredit a physiognomic account of expression and put in its place a scientific understanding of expression based on physiological ideas of its function.[15] Physiognomy toed an orthodox religious line, disseminating a theological world view in which the appearance of things was ultimately taken as a sign that the creator was active in the world; theories of expression based on physiology, on the other hand, presented a more heterodox view of the organic world in so far as they stressed the integration of mind into body. The latter reflects the growing importance of materialism, emerging first in this study in the work of David Hartley followed by Charles Bell; if accepted – and it was not always a straightforward acceptance, as the examples of Hartley and Bell show (compared to Bain and Spencer, for instance) – theories of matter worked against the idea that we can infer an inner meaning from the appearances of things in the world, and proposed instead an understanding of expression conceived in terms of physiological function, in particular at a neural or muscular level.

That said, the distinction between orthodox and heterodox views of the world requires clarification if it is to be used in the context of nineteenth-century scientific thought. Historical writing on nine-teenth-century science has changed radically over the past thirty years. Whiggish accounts of scientific discovery and progress have been replaced by a contextualising approach which considers the network of institutions and affiliations, theories and practices within which any claims for science (or indeed any epistemological claims) should be discussed. Thomas Kuhn's account of the development of normal science – or work which uses existing theories to predict certain factual outcomes – as the result of a series of intellectual revolutions has strongly influenced this change of direction in the history of science.[16] He claimed that our image of science as *the* paradigm of rationality embedded in specific institutions is seriously distorted; instead of viewing the scientist wielding his scientific method on a path towards truth, he proposed a characterisation of science which takes account of the social realities and cultural pressures within which scientific practice is embedded.[17] It is against this background that recent work in the history of science has helped us appreciate the fluid and often loose definitions of science and scientific communities.[18] 'Early Victorian science was volatile and underdetermined', Alison Winter has argued, 'people could not agree about what one could safely claim about natural law, nor was it obvious when, where, and to whom such claims could be made'.[19] It is into this context of an open and heterogeneous scientific community that physiognomy should be placed: admittedly, it is a discipline with an unusual constituency, but it does involve a loose definition of science which reflects the wider debates about the status of science and the claims made for it.

Physiognomy was presented as a science of mind designed to reveal the moral order: it was, Lavater believed, an orthodox science and yet it was rejected by sceptics on the grounds that it was a profoundly unorthodox version of a science of mind.[20] The point is that the orthodoxy of physiognomy within a religious framework was not sufficient to guarantee the orthodoxy of physiognomy within a scientific one. Yet the physiognomic practice of the nineteenth

century was founded on a double-edged appeal, towards religion on
the one hand and science on the other. Placed alongside the scientific
projects of, say, geology, physiology, and biology, it is impossible to
make physiognomy stand up in equivalent terms, but if we place it
alongside the projects of phrenology and mesmerism, for instance,
physiognomy starts to look more plausible as a form of what has
been variously termed 'pseudo'-science, 'alternative' science, or even
'quackery'.[21] The problem is that characterisations of phrenology,
mesmerism, and physiognomy as 'pseudo' or 'alternative' forms of
scientific inquiry presuppose that we can identify claims about the
organic world as falling on one side or other of a boundary line
which demarcates what is science from what is not, and in this way
define the realm of science proper. However, as Winter implies, the
notion of a boundary line between science and non-science has the
effect of smoothing over the diversity of opinion in the scientific
communities of the nineteenth century and positing in its stead a
rather singular and monolithic view of science – a view which
scholarship on the period has now affirmed was not evident. 'If
proper science could be defined differently in different contexts',
Winter concludes, 'then scientific claims could have radically different status and even, perhaps, different meanings depending on
where they were read or heard and by whom'.[22] The question we
should ask of physiognomy, then, is not whether it counts as proper
science but what kind of claims it makes about its practice as
scientific. In other words, because it defines itself as a science of
mind, albeit a popular and subjectively grounded science, physiognomy raises important questions about the cultural and epistemological status of science.

Though nineteenth-century discussions of science and scientific
method involved some reference to the logical process of scientific
discovery, there were some basic epistemological questions about
reality which could not be easily dealt with. These included questions on the relationship between observer and observed, the value
of scientific knowledge in relation to more intuitive forms of
knowledge, and the role of science in society. It is widely thought
that William Whewell's pronouncement in his *Philosophy of the
Inductive Sciences* (1840) on the need for a name to describe 'a
cultivator of science in general' brought the very word 'scientist' into
common usage as a term of description.[23] In fact, Whewell had used
the term a few years earlier to describe a heated discussion which

had occurred at the newly founded British Association for the Advancement of Science in Cambridge in 1833:

Philosophers was felt to be too wide and too lofty a term, and was very properly forbidden them by Mr. Coleridge, both in his capacity of philologer and metaphysician; savans was rather assuming, besides being French instead of English; some ingenious gentleman proposed that, by analogy with artist, they might form scientist, and added that there could be no scruple in making free with this termination when we have such words as sciolist, economist, and atheist – but this was not generally palatable.[24]

The sense of this pronouncement is relatively clear – that is to say, there is an activity associated with science which is, fundamentally, a human activity – but it is less obvious here whether the activity of science was, in Whewell's opinion, designed to reflect and reveal the omnipotence of the Creator.[25] An admirable apologist for science, Whewell seemed to appreciate that the task of defining science and the activity of scientists was a cultural and philosophical problem which required considerable thought about the process and end of science. 'It is no easy matter, if it be possible', he wrote, 'to analyse the process of thought by which laws of nature have thus been discovered'.[26] The use of 'process' rather than 'method' is significant here because it suggests the suppleness (or some may say looseness) of Whewell's understanding of science: it represents an attempt to construct a theory of science which drew on both personal experience and rational thought. Such a conception of science was heavily dependent on induction as the mediating link between experience and thought or, as he put it, 'observation of Things without, and in an inward effort of Thought; or, in other words, Sense and Reason'. 'The impressions of sense', he went on, 'unconnected by some rational and speculative principle, can only end in a practical acquaintance with individual objects . . . [whereas] the operations of the rational faculties . . . if allowed to go on without a constant reference to external things, can lead only to empty abstraction and barren ingenuity'.[27] The assumption is that sense and reason can co-exist provided there is a place for personal experience in our understanding of science.

Claims about the status of scientific knowledge originate, according to Whewell, from an individual's highly subjective view of the world and become a theoretical construct when the philosopher or scientist recognises the 'Fundamental Idea' which explains observed phenomena. Truth as such, inductive truth, is not discovered but intuited over a period of time as a result of a prolonged and active engagement between the conceptual framework of the individual and the objects of the external world. 'There are scientific truths which can be seen by intuition', Whewell explained, 'but this intuition is progressive'.[28] In other words, science makes use of Fundamental Ideas to describe the objects of the external world, but because scientfic investigation is an essentially human activity, and truth is a metaphysical form which defies absolute definition, there will always be a residue of truth in the Idea:

The Idea is disclosed but not fully revealed, imparted but not transfused, by the use we make of it in science. When we have taken from the foundation so much as serves our purpose, there still remains behind a deep well of truth, which we have not exhausted, and which we may easily believe to be inexhaustible.[29]

The whole pattern of Whewell's process of induction – which resembles what is now called the hypothetico-deductive character of developed sciences such as physics or biology – depended upon the emergence of Fundamental Ideas which divulged rather than fully grasped truth, and so presented science as a system of successive generalisations of observed particulars.

The key question asked by Whewell seems to be the following: how do we evaluate claims to knowledge which are at once subjective and theoretical? What is interesting in his reply is the role he accords to intuition in the process of scientific thought, particularly as intuition has an ambiguous philosophical history. Intuition is often taken to mean something like the capacity to arrive at decisions or conclusions without the benefit of conscious, rational thought processes. In its philosophical sense, it has been used to denote the alleged power of the mind to perceive or grasp certain self-evident truths, but its status as a mental process has paled into insignificance next to the analytical rigours of formal logic. A robust rejection of the place of intuition in scientific theorising can be found in the arguments of Lewis Wolpert who claims that science must be considered an 'unnatural' mode of thought. Wolpert's thesis is predicated on a simple binary opposition between what is natural (common sense) and what is unnatural (science). To accept the unnaturalness of science is, according to Wolpert, the first and necessary step towards correcting many of the misunderstandings about science and the status of scientific knowledge: 'doing science

. . . requires one to remove oneself from one's personal experience and to try to understand phenomena not directly affecting one's day-to-day life, one's personal constructs'.[30] The obvious point here is that the conditions necessary to everyday thought are not sufficient for scientific thought, or in other words intuition is not and cannot be the basis for scientific thought. However, John Dupré has argued to the contrary that the foundation of science on metaphysics contributes to the dream-like fantasy that science can give us an orderly and complete account of things in the world. Such a dream involves an ideal of scientific unity based on a set of assumptions which, Dupré claims, are undermined by their own conclusions. In rejecting these assumptions – the doctrines of essentialism, determinism, and reductionism – Dupré denies that science comprises a single, unified project and instead proposes a realist philosophy of science based on epistemological pluralism, or a 'promiscuous realism' based on 'the disorder of things'.[31] It is a compelling argument about the nature and extent of the claims which can be made for the activity of science in the world, pointing out that ordinary language (not science) proffers the basic model from which scientific taxonomies are constructed.[32]

I argue, following Dupré, that the process which instinct (and often intuition also) describes – our ability to grasp things without being aware of how we do so – underpins almost all of our everyday actions, judgements, and decisions. Equally, it has a significant part to play in nineteenth-century debates about scientific knowledge. What we find in physiognomy is the mediation of these two aspects of instinct, as an integral part of everyday life and an important element of scientific theorising. Lavater's claim of scientific status for physiognomy relies on instinct, which derives meanings also from contemporary moral and social language. It is in relation to instinct that Lavater can make physiognomy the basis of all human actions and, conversely, that Darwin radically overhauls physiognomy in favour of a theory of expression as a physiological scientific statement. Yet, as we will see throughout this book, questions about the rationale for theories of expression based on physiology are bound up, almost inevitably, with questions about the status of the scientific knowledge they involve. Is it possible, for instance, for knowledge gleaned from expression to be scientific? What would it take for it to be so? I want to suggest that knowledge of expression was normative and empirical in content whether in respect of physiognomy or physiology. As a result, the changes in the understanding of what was involved in expression which form the subject matter of this book advanced a study of science as one dimension of a shared culture.

The claim by physiognomy to be a science of mind was central to this notion of a shared culture, as it rendered expression a product of the mind which was in turn the reflection of a higher being. If the appeal of physiognomy lay in its reliance on largely instinctive responses to external appearances, its purpose was to identify and to describe the common forms which organised the diversity of appearances. In this way, as I have suggested, physiognomy drew on the moral and social language of the day in order to guarantee the claims it made about human nature. The knowledge of expression, emotion, and character derived from physiognomy was predicated on practical attitudes and widely held beliefs which, though hard to justify, were generally held to be reliable and, moreover, it functioned in a profoundly normative manner as the determinant of that which was shared by all people and things in the organic world. Man was the centre of all things, and all his works – body and mind, culture and society – were, in principle, the means of understanding the organic world. As Roger Smith has demonstrated, such theories of man and mind produced by the natural philosophies of the seventeenth and eighteenth centuries tended to persist in using ideas of purpose, power, and action to express the active part of physical events.[33] Given that expression was arguably one of the most explicit of the physical states in the world, a form of mediation between a natural and a *super*natural order, the actions of a higher mind could be comprehended through the various kinds of facial expression. Hence, each and every attempt to read and judge character was a means of ascribing an essence to human nature – a form of imagining there was something hidden from external appearances which, once revealed, made them more purposeful and more substantial.

These are the conditions within which the emergence of theories of expression based on physiology should be considered. The first half of the book establishes the significance of the physiognomical tradition as a framework for the physiological study of expression. Chapter one considers the essentialist argument which lies behind physiognomy and reflects on the contribution Lavater's physiognomical teachings made to debates in late eighteenth-century natural philosophy about the nature of man, and in particular about the

human mind. David Hartley was largely responsible for the contentious nature of theories about the association of ideas and the
physical basis of the mind. He proposed a model of mind that sought
to bring together mind and body in an integrated system which was
regulated solely by vibrations and associations, primarily those of the
sensory apparatus. It was a sophisticated approach to the problem of
mind which synthesised the most recent contemporary ideas of
mental processes. A comparison of Lavater's conception of human
nature and the work of Hartley (together with the earlier, explicitly
Cartesian ideas of Charles Le Brun) will give us an understanding of
how Lavater's largely instinctive insights into expression might
produce a physiological explanation of emotion, instinct, and sensation.

Chapter two explores the claims and counter-claims of Charles
Bell and his contemporaries for the kind of conceptual framework
which might organise nature and the conditions of existence. To
Bell, there was no question that the concept of design was the means
of making sense of nature, but his problem was how to reconcile his
investment in natural theology with a developing interest in the
structure and function of the nervous system. Whilst Bell's work is
important because it establishes the empirical grounds upon which
physiognomical teachings can be verified, it is also notable because it
signals the importance of the new physiological doctrines in establishing functional explanations for the actions of the nerves. What
we then see in chapter three, is the extent to which Bell's treatise on
expression had important ramifications for the artists of the Pre-
Raphaelite Brotherhood (PRB) who resolved to record and represent
the everyday world, the world of the senses, through a very
particular notion of character. An expression of character must, for
the PRB, involve capturing a truth which comes from the direct
observation of life; that is to say, life as dramatic moments of
excitement and understood as inner spiritual and moral experience.

The second half of the book examines the rise and triumph of new
physiological doctrines which provide inner, scientific rationales for
the expression of the emotions. Chapter four examines the connection between appearance, character, and intellect, and in particular
the appropriation of the language of expression to convey certain
culturally loaded meanings about the virtue of healthy physiques
and the moral worth of beauty. While Hartley's associationist ideas
emerge most clearly in the work of Alexander Bain, for whom the

will and all experiences and affections are resolved into sensations and ideas, we see in Herbert Spencer's studies the espousal of a new evolutionary fascination with healthy stocks and perfect forms. Chapter five develops this discussion through an investigation of Darwin's study of expression and emotion in man and animals. The beginnings of a science of human character and behaviour in the final quarter of the nineteenth century are certainly due partly (though not solely) to Darwin, and yet his study of expression makes a number of contradictory claims about the possibility and plausibility of conducting a scientific analysis of expression. Darwin presents his study as the cornerstone of his evolutionary theory, but there seems to be little in the way of guarantees from evolutionary theory that genealogy can provide us with the distinctions we need in order to understand the products of evolution as well as the process by which they came to be. Hence, what results is a study of expression which displays rather than dismisses elements of the practice of physiognomy.

The final chapter looks forward to Francis Galton's ideas on resemblance, hereditary transmission, and the unity of type. It is clear that Galton takes physiognomy to its logical conclusion in so far as he deems the survival of the species to be dependent on the selection of its healthiest and most agreeable types; that is to say, those proper to the advancement of the natural order. In effect, Galton constructs a system of judgement which allows the practical application of physiognomic ideas to assume a central role in ordering and regulating society by emphasising human action. Thus, the movement from a physiognomic account of expression to a physiological one can be described in terms of the process of replacing the idea of mind as a spiritual essence of a *super*natural order with a physicalist conception of mental states. At the same time, however, the development of a naturalised conception of human nature (such as that proposed by the writers discussed in the second half of this book) requires physiognomy as its anchor – the evolutionary explanation of expression is, then, the long-term outcome of physiognomical teachings as well as the reason for their dissolution.

CHAPTER I

A science of mind? Theories of nature, theories of man

To divide and arrange the body into organs, and to ascribe to each its functions, is physiology. To view all these organs in connexion, and to compute the influence of each, and the concentrated influence of the whole, in determining the great movements of the individual among other individuals, all acting their respective parts in the great struggle and bustle of life, is physiognomy. Physiognomy is just a system of corollaries arising out of physiology.

John Cross[1]

I

In 1746, James Parsons gave the Crounian lectures to the Royal Society on the subject of 'Human Physiognomy Explain'd'.[2] The lectures were intended to demonstrate the place of physiology in explaining muscular activity and so provide a common context for the discussion of ideas about the structure and function of living organisms. These discussions were essentially about the relationship of mind and body and the extent to which agency could be ascribed to the mind in the body; at issue was the preservation of the independent existence of the soul and the distinctions conferred on life and mind as a result, including the establishment of a barrier between man and all other organisms. Physiological principles, such as that of the reflex action, the nerve function, and the notion of the *sensorium commune*, which were implicated in a view of the individual as controlled by mind, were gradually reworked from the eighteenth into the nineteenth century to support an understanding of the command of the nervous system over the individual.[3] To describe human nature, then, was to become enmeshed in often very detailed arguments about the actions of the nerves and the muscles, the purpose of sensations and feelings or emotions, and the efficacy of the will. As Roger Smith has claimed, 'if we include ancient beliefs in the humours and in sympathy, [physiology] had long been central in guides to well-being'.[4]

By drawing a correspondence between physiological actions and physical expressions, Parsons presented an account of physiognomy which demonstrated the importance of the mind to an understanding of the expression of the emotions. 'I shall now attempt to give you a Description of the Muscles of the Face', he wrote, 'with some Observations and Remarks, which I hope will appear curious to you, relating to their separate as well as conjunct Actions, and the Appearances of the Countenance that are the natural Effects of such Actions'.[5] Though the main tenet of physiognomy was that the physical features of a person's face and body indicated character, instincts, and behaviour, and all these were expressive of the soul, Parsons was mainly concerned with the muscles of the face, the purpose of which was to preserve the moral disposition of an individual. As he claimed, 'the several Motions of the Face that express the different Passions of the Mind . . . serve two principal Ends':

first (altogether) to form the Symmetry of the Countenance, by supporting the Skin of the Face, in the Manner we see it when a general *Composure* appears thro' the Whole; and, secondly, to express, as we have said, those Passions of *Joy, Grief, Fury, Ill-nature*, and such-like, as the Mind is often prone to suggest.[6]

Emotional expression was, to Parsons, the product of the nervous system and in particular the diaphragm, which acted as the instrument of the will (or mind) in conveying some of its impulses and impressions to the face. He insisted:

It was because the Means of Self-Preservation should be generously distributed to us, that the prevailing Characteristics of Tempers should be thus conspicuous in us; innumerable Instances of which are to be observed in every other Part of the animal World besides: And even from hence we might naturally conclude it absolutely necessary; but the Structures of these Parts, their sensible Actions, and the great Consent between one Part of the Animal and another (from their nervous Communications), yet more plainly confirm this Conclusion.[7]

The nervous sympathy between the diaphragm and the muscles of the face provided a means of communication which seemed to involve the transmission of impressions along the nerves to the face; it was this that Parsons termed 'nervous communications'. As a result, the expressions of the emotions are made evident on the faces of man and animals and serve to preserve, and, it is implied, to control, the identity and individuality of creatures in the world.

Parsons used his two lectures to trace the history of physiognomy and illustrate the principles of its practice, drawing especially on the work of Aristotle, Charles Le Brun, and Giovanni Battista della Porta. The lectures were mostly summary in form but they were important in identifying the main principle of physiognomy via the habitual action of the muscles which were expressive on the face. 'I hope it will appear', he explained,

> that no Analogy can be drawn from Brutes, no Signs from the Voice, nor general Shape of the Face, or any of its Parts; in a Word, nothing but the Actions of the Muscles, become habitual in Obedience to the reigning Tempers of the Mind, can in any wise account for them; and the Art of Physiognomy, especially the *Metoposcopy*, or what relates to the Face, must prove very uncertain without this Foundation.[8]

To claim that muscular action was directly related to mental processes, and that this relationship was repeated so often it became habitual, was more consistent with the physiological principles of the second half of the eighteenth century than with its physiognomic ideals. Of all the general philosophy about nature and man at this time, physiology was that which was concerned with the relationship between mind and body, as I have suggested. The emergence of experimental methods and practical techniques in the physiology and medicine of the nineteenth century tended to promote a conception of human existence in primarily physical terms, but the idea that the mind was dependent on organised physiological structures, or in other words matter, was (and remained) a contentious issue.[9] The interest of Parsons's lectures in this context was his readiness to conceptualise physiognomy in loosely physiological terms, and in so doing, to allude to the discussions about human nature which were appearing at the time. Whilst Parsons does not go so far as to claim a material basis for mind, he is, quite clearly, suggesting a link between mental and physiological actions, in particular as exemplified by the expression of the emotions which physiognomy describes.

More than seventy years after Parsons gave the Crounian lectures, John Cross aligned physiognomy and physiology in a similar way, as the opening quotation to this chapter shows. Physiology was, to Cross, the division of the body according to its organs and the classification of those organs via function, whereas physiognomy is the observation of these organs as a unified whole which effect the action of an individual. The implication is that physiology and physiognomy shared a common concern in investigating the nature of life, as Cross intimates that whereas the former looks to an understanding of structure and function to explain action, the latter takes action as evidence of a complex internal organisation. Hence, 'physiognomy is just a system of corollaries arising out of physiology'. Parsons' statement is both unequivocal and shrewd since what prevented physiognomy from becoming widely accepted as a science of mind was its inability to construct a model of mind which explained the correlation of mental and physical activity. Instead, physiognomy assumed that life, and especially the life of the mind, was explicable only by referring to the nature of the soul and the relation of man to God. For Parsons and Cross, the alignment of physiognomy with physiology was appealing because it allowed the suggestion of a theory of life based on physical principles without it being wholly incompatible with a belief in the transcendental power of mind.

It was between Parsons and Cross that physiognomy became popular once more, revived and supported by the publication of Lavater's *Essays on Physiognomy*, initially successful in German, for which there were five editions in the 1770s and four in the 1780s, then in French in the 1780s, followed later in the same decade by two English editions.[10] Lavater presented physiognomy as a science of mind which construed human actions as forms of moral behaviour, and, as suggested in the introduction, the practice of physiognomy drew primarily on theological notions but also on physiological ideas of man and nature in order to make its points. As L. S. Jacyna has argued, physiology 'passed easily into other areas of natural philosophy and also into the domain of morals and religion':

Theories about the body did not constitute a discrete sphere relevant only to a few professionals; rather they were an aspect of a common stock of ideas that could be drawn upon for a variety of purposes. In particular, physiological notions remained part of the currency in which the commerce between social groups was conducted.[11]

The alignment of physiognomy with physiology made by Parsons and Cross was not, then, coincidental but rather indicative of the contradictions involved in discussions of human nature, and indeed seems to represent an attempt to include physiognomy within the remit of the sciences of mind. The big question was whether man was separate from nature and the laws of the organic world or integrated into that world. This chapter considers that question, elaborating on the relationship between mind, metaphysics, and expression, and showing the extent to which Lavater's ideas fitted into debates about the place of man in nature. A comparison of Lavater's conception of man and the earlier work of Charles Le Brun and David Hartley will suggest how Lavater's largely intuitive insights into expression might be compatible with a natural scientific study of emotion, instinct, and sensation. I shall therefore sketch the different models of mind presented by Le Brun and Hartley before discussing in some detail Lavater's conception of man, particularly in the light of the physiological context which the earlier writers provide for Lavater's ideas of expression.

II

Charles Le Brun's lectures to the Académie de Peinture, founded in Paris in 1648, addressed the nature of expression as a symbolic form of language with a specific relevance to painting:

Expression, in my opinion, is a simple and natural image of the thing we wish to represent; it is a necessary ingredient of all the parts of painting, and without it no picture can be perfect; it is this which indicates the true character of each object; it is by this means that the different natures of bodies are distinguished, that figures seem to have movement, and everything which is imitated appears to be real.[12]

Three lectures were planned – on expression in painting, a theory of expression, and a system of physiognomy – and, as Jennifer Montagu has shown, it is unclear not only when these lectures were delivered but also whether all of them were given.[13] What is clear, however, is that a lecture on expression, *Conférence sur l'expression générale et particulière*, was delivered by Le Brun to the Academy in 1668, and that its popularity was such that it appeared in more than sixty editions throughout the next century. Le Brun's goal was to instruct the artist in the workings of expression by presenting a systematic account of emotional expression based on physiological principles, and in so doing he drew heavily on the philosophical and physiological writings of René Descartes. What he found helpful in Cartesian thought was the extent to which Descartes' theory of the relation of mind to body relied on the legibility of the passions through the actions of the body and as expressive of the mind (soul). As a result, the notable aspect of Le Brun's work was his attempt to apply the method of deduction to the study of expression with the consequence that the process of observation was a secondary matter when compared to the primary process of deductive reasoning from *a priori* rules.

Expression was, to the artists of the seventeenth and eighteenth centuries, a study of the passions represented through the gestures, features, and movements of the face and body. The following example, a definition of the term given by Roger de Piles, prominent art theorist and contemporary of Le Brun's, indicates the complexity of its application to painting:

Expression, when speaking of painting, is completely confused with *passion*. They differ, however, in that *expression* is a general term which signifies the representation of an object according to the character of its own nature, as well as the particular emphasis the painter has designed to give it for the purposes of his work. *Passion*, in terms of painting, is a movement of the body together with certain features of the face, marking some agitation of the soul. It follows that every *passion* is an *expression*, but every *expression* is not a *passion*. There is no object in a painting that does not possess its own expression.[14]

The challenge that de Piles sought to resolve was how to distinguish individual expressions from general types of expression, and then visualise how these types function. To paraphrase de Piles, expression operates on two levels: on the one hand, it is the general name used to describe the nature and character of an object, and on the other, it is the particular impression given to that object. Passion is the action or emotion which precedes and causes the expression, so even though every passion corresponds to a specific kind of expression, there is no guarantee that everything with an external expression will convey passion. Paintings may depict facial expression but they can still be empty of passion. As de Piles insisted, each and every object represented in the visual arts has its own expression but not all of these objects could be deemed to indicate passion and emotion. The distinction is significant in so far as every form represented in painting has a particular expression, but only those forms which convey passion can have a general, perhaps even universal, expression.

Like De Piles, Le Brun sought an understanding of expression as a key to discerning the actions of the soul. He explained:

First, a passion is a movement of the sensitive part of the soul, which is designed to pursue that which the soul thinks to be for its good, or to avoid that which it believes to be hurtful to itself. Ordinarily, anything which causes a passion in the soul produces some action in the body.[15]

To put it in other words, an emotion is primarily the product of the mind and causes a reaction in the body, the nature of which is dependent on self-preservation. This idea of the connection of mind or soul to body was, of course, the fundamental precept of Cartesian thought and was borrowed by Le Brun as the rational foundation for his theory of expression.[16] The phenomena of the mind (soul) were, for Descartes, independent of the phenomena of the physical world and had, instead, a completely autonomous status; hence he claimed in *Discourse on Method* (1637): 'I knew I was a substance the whole essence or nature is simply to think, and which does not require any place, or depend upon any material thing, in order to exist'.[17] He maintained, in fact, the existence of two radically different kinds of substance, a physical, extended substance (*res extensa*) and a thinking substance (*res cogitans*), of which the first has length, breadth, and depth and so can be measured and divided whereas the second is unextended and indivisible. On this basis, the human body (including the brain and the nervous system) is categorised as a physical substance and the mind (including thoughts, desires, and volition) is categorised as a non-physical substance.[18] This dualist view makes the first task of the philosopher a study of the mind (soul) which must be regarded as prior to nature and irreducible to matter. Thus, as Descartes suggested in a later work, *The Passions of the Soul* (1649), even though the mind was a non-physical entity, it had the capacity to exercise its functions in a central part of the brain – namely, the pineal gland:

What is a passion in the soul is usually an action in the body . . . [and] anything we experience as being in us, and which we see can also exist in wholly inanimate bodies, must be attributed only to our body. On the other hand, anything in us which we cannot conceive in any way as capable of belonging to a body must be attributed to our soul.[19]

The passions were affections of the soul that functioned through the pineal gland which, in turn, regulated the responses of the body and influenced the flow of what were termed spirits to the muscles.

To all intents and purposes, Le Brun's theory of expression simply restates these Cartesian ideas: expression, principally of the face, provided a series of patterns for understanding how the mind (soul)

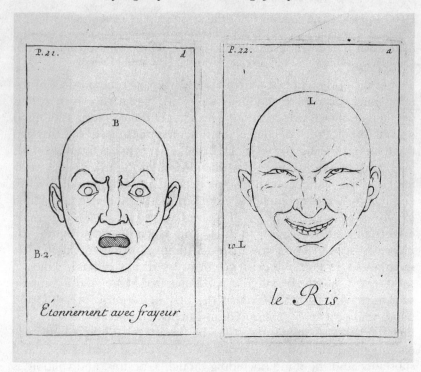

Plate 1   Charles Le Brun, Two outlines of faces showing astonishment and fear
(left) and laughter (right), etching by B. Picart, 1713.

was active in the physical world, not because the face was a physical
entity but because it was proximate to the brain and so believed to
be the most accurate index of the mind; at least its features, and in
particular the eyebrows, were thought to be so (plate 1). A knowledge
of these principles, philosophical and physiological, which directed
the activity of the mind and body would, he claimed, release the
artist from simply copying nature and allow him to create his own
images directed by and perhaps even improving on the processes of
nature.[20] Le Brun's understanding of expression, as stated in the
Conférence sur l'expression, was based around three areas of research –
the heads of ancient rulers and philosophers, specific studies of the
eyes of men and animals, and a comparison of the heads of men and
animals (plate 2)[21] – and his task was to demonstrate the correlation
between the expressions of the face, its muscular action, and the
passion or emotion which causes both action and expression. 'An
action is nothing else but the movement of some part', he wrote,

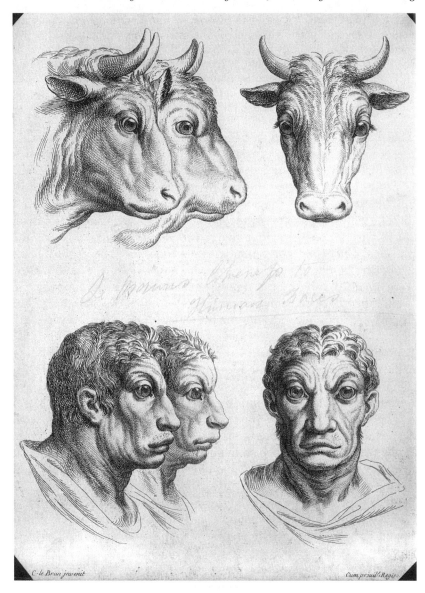

Plate 2   Charles Le Brun, Three perspectives on the head of an ox and three on the head of an ox-like man showing the physiognomical relations between certain members of the species, etching *c.* 1820.

and this movement can be effected only by an alteration in the muscles, while the muscles are moved only by the intervention of the nerves, which bind the parts of the body and pass through them. The nerves work only by the spirits which are contained in the cavities of the brain, and the brain receives the spirits only from the blood which passes continuously through the heart, which heats it and rarefies it in such a way that it produces a thin air or spirit, which rises to the brain and fills its cavities.[22]

This vision of the interrelation of body and mind relies on a mechanistic chain of increasing complexity, which works backwards from the movement of a part of the body, and a corresponding response in the muscles and the nerves, to the circulation of spirits and blood and their influence on the brain. By amplifying Descartes' division of the passions into simple and mixed kinds, Le Brun worked from the basis that there were four characteristics of the passions which had corresponding movements of the eyebrows; so whilst the simple passions of love, hatred, desire, joy and sorrow were made manifest in a movement 'which rises up towards the brain', the mixed passions of fear, courage, hope, despair, anger, and fright were manifest in a movement 'which slopes down towards the heart'. 'In proportion as these passions change their nature', Le Brun said, 'the movement of the eyebrow changes its form, for to express a simple passion the movement is simple, and, if it is mixed, the movement is also mixed; if the passion is gentle, so is the movement, and if it is violent, the movement is violent'.[23]

The point is that the physical form of the body can be seen, according to Le Brun, to support a non-physical conception of mind. Le Brun argued that each individual had a dominant sign or facial feature which revealed their character, based on the *a priori* fact of the existence of soul. This feature, the slope of the eyes, worked in tandem with the movements of the eyebrows to indicate the kind of character under analysis (plate 3). So, for example, eyes which sloped upwards suggested to Le Brun a 'spiritual' kind of character; eyes which were level suggested a 'normal' kind of character; and eyes which sloped downwards suggested a 'base' kind of character. By dividing characters into kinds or types in this way, Le Brun established a system which assumed there was a hierarchy of characters among mankind. Having adopted a Cartesian model of mind, Le Brun's metaphysical system of expression had emotional states (or states of passion) acting as illustrations of the mind – as the physical means of comprehending a non-physical entity – and in

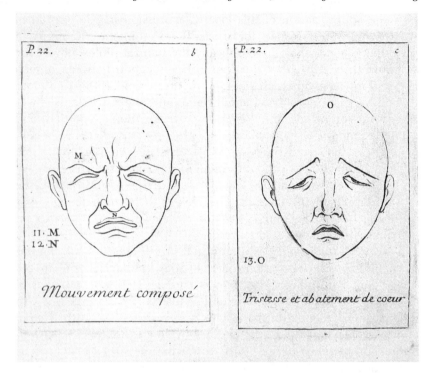

P.22.   *b*

M

11. M
12. N

Mouvement composé

P.22.   *c*

O

13. O

Tristesse et abatement de coeur

Plate 3   Charles Le Brun, Outlines of faces expressing emotions (left) and sadness
and dejection (right), etching by B. Picart, 1713.

effect rendered each expression a metaphor of mind, and each idea
of mind a source of knowledge about the soul. His object was to
demonstrate precisely how and why this metaphor worked, and he
did so by using the eyebrows as patterns for understanding the
activity of the mind in the physical world.

A leading feature of Le Brun's understanding of expression was,
therefore, that the action associated with facial expression, and in
particular that of the eyebrows, was nothing more than the movement of a part which, like the dial of a clock, concealed a complex
mechanism behind its external appearance (see plate 1).[24] There is
a link, at least implied, between the fact of this action and what we
might call an axis of pleasure and pain. Le Brun's idea of the soul
differentiating between what is 'for its good' and what is 'hurtful to
itself' provides a means through which expression can function as a
human act which informs moral behaviour, and it is easy to see

how the appropriation of this idea of expression in painterly form was an appealing prospect for artists. It is in something more like a scientific context, though, specifically the natural philosophy of the seventeenth century, that this sense of connection between human acts and moral conduct assumes an important role in thinking more carefully and in more detail about the nature of action. At issue was the distinction of voluntary as opposed to involuntary motion, and the participation of the soul, along with consciousness and the will, in creating such a distinction. As we shall see later in this book, debates about the position of the soul in respect of the voluntariness or involuntariness of action remained contentious until nearly the end of the nineteenth century. The crux of the matter, though, was the capacity of an immaterial form, like the soul, to explain involuntary or reflex actions.

Descartes' dualistic thesis, predicated on the mechanical motions of man's material body controlled by the immaterial soul in the pineal gland, introduced the concept of reflex action as a means of describing the action of the soul and its interaction with the body. In their impressive book on the origins of neuroscience, Edwin Clarke and L. S. Jacyna explain the action of reflex as follows: 'a sensory impression travelled to the brain, from where it was reflected . . . as in the manner of light, into motor nerves to bring about muscular contraction'.[25] This description makes clear the division of the physical from the mental realm and points to the difficulty with Le Brun's idea of mind; namely, his adherence to the idea of a mind placed outside the organic world which is visible only indirectly through the motions of expressions. His study of expression represents an attempt to mesh a *super*natural theory of mind with a description of its structure or function in the physical world, and as such its weakness was in the sketchy physiological explanations it offered for mental processes whereby as long as the mind remained outside the world, any knowledge of its physiological structure or function remained outside the remit of enquiry (and so extremely tentative). Though Descartes had provided a deterministic model in which purposive actions could be explained through law-like processes – and Le Brun had largely followed Descartes' lead – there was increasing interest in the seventeenth and eighteenth centuries in the notion of reflex, and numerous attempts at conceptualising the causal process linking sensory impressions with mental and motor responses emerged to counter Cartesian doctrine. How could it be that removing the head from an animal, for instance, did not cut out all motion from its body? Many names could be cited here – Stephen Hales, Thomas Willis, Robert Whytt, for example – but it is generally held that the combined efforts of Isaac Newton and John Locke did most to loosen the hold of the Cartesian model of mind on the materiality of the body.

This is the backdrop against which David Hartley, British philosopher and physician of the eighteenth century, developed an account of mind based on the capacity of the nervous system to respond to stimuli through a series of vibrations transmitted through the body and associated in the mind to produce certain ideas. Following Locke, philosophies of nature in the eighteenth century made an increasing number of speculative inquiries into the correspondence between physical objects and mental ideas, proposing a relationship between the physical and the mental realms in terms of their respective functions.[26] For Hartley, though, the goal of this kind of inquiry was the removal of a metaphysical conception of mind (soul) in favour of a naturalistic understanding of body and mind. In particular, he suggested that the common-sense experience of an organism (integrating body and mind) should be the focus of attention. The physiological psychology connected with Hartley surfaced quite clearly from the mind/body problem, for it denied the conception of mind as the property or essence of a supernatural order and instead offered an account of the interdependence of mental and bodily factors which became one of the most systematic (and speculative) investigations into the physical constitution and properties of man. Le Brun's work on the passions represents the first attempt at constructing a scientific explanation of the expression of emotions but, in my opinion, the real foundations for such an explanation are laid with Hartley's understanding of the physical relationship between sensations and ideas, and the resulting mental processes, which strikes at the heart of dualistic notions not only of mind and body, as I have indicated, but also of reflex action.

III

'Man consists of two parts, body and mind', Hartley wrote in the opening lines of *Observations on Man* (1749): 'the first is subjected to our senses and inquiries, in the same manner as the other parts of the external material world [and] the last is that substance, agent, principle, &c. to which we refer the sensations, ideas, pleasures, pains, and voluntary motions'. He continued, 'sensations are those internal feelings of the mind, which arise from the impressions made by the external objects upon the several parts of our bodies' and ideas are all other 'internal feelings'.[27] Hartley proposed the reduction of mind to the sensory-motor functions of the body; hence, normative concepts of mind (such as reason, memory, and will) and concepts which determine character and personality (such as sensations, ideas, and muscular motions) were, he claimed, explicable through the doctrines of vibration and association. The nervous system was the centre of Hartley's investigation, and by this time, as Karl Figlio has argued, it had come to represent 'the bridge between the philosophical/psychological inquiry into the soul and the nature of man on the one hand, and the anatomical/physiological study of their structure and function on the other'. Figlio's identification of two kinds of inquiry is significant because, he claims, they stand for the distinction between mind and matter of Cartesian dualism: there is on the one hand, then, a philosophical inquiry into the mind and on the other a physiological inquiry into the body.[28] Hartley's explanation of mind attempted to unite both these kinds of inquiry by taking the integration of mind into body as the basis for a functional description of human nature.

Central to Hartley's thesis was the notion that human acts arose out of pleasure or pain, of which there were seven classes: sensation, imagination, ambition, self-interest, sympathy, theopathy, and the moral sense. The response of an individual to a specific stimulus was determined by the transmission of impressions through the nerves and into the brain, which worked rather like a musical instrument in so far as it vibrated upon receiving the impressions of external objects. It was the reception of these impressions in the brain, (composed of the cerebrum, the cerebellum, and the medulla oblongata) as internal feelings that Hartley termed ideas of sensations (simple ideas), and they were preliminary to the emergence of intellectual ideas (complex ideas) that produced voluntary motions as opposed to the involuntary motions of an organ like the heart. He explained:

External objects, being corporeal, can act upon the nerves and brain, which are also corporeal, by nothing but impressing motion on them. A vibrating motion may continue for a short time in the small medullary particles of the nerves and brain, without disturbing them, and after a short time would cease; and so would correspond to the above-mentioned short continuance of the sensations; and there seems to be no other species of motion that can correspond thereto.[29]

The claim was that the frequent repetition of the vibrations of sensation caused vibrations of a lesser force in the medullary substance of the brain; and these so-called after-vibrations enabled ideas to be linked or associated with each other and in so doing form clusters and combinations of simple ideas (of sensation) which coalesced into complex ideas (of intellect). 'Let us suppose', Hartley wrote,

the first object to impress the vibrations A and then to be removed. It is evident from the nature of vibratory motions, that the medullary substance will not, immediately upon the removal of this object, return to its natural state N, but will remain, for a short space of time, in the preternatural state A, and pass gradually from A to N.[30]

Through sustained action of this kind the natural state gives way to the preternatural state and becomes the feeding ground for the association of various immediate and then delayed vibrations.

Borrowing the Lockean understanding of the mind as a blank state or *tabula rasa*, Hartley proclaimed that in the early stages of life simple ideas of sensation predominated, whereas the recurrence of these ideas from childhood into adulthood brought with it the capacity to associate simple ideas and so ensure that complex ideas dominated. In effect, what Hartley was positing was a developmental model of the interrelation of mind and body which became increasingly sophisticated through life in dealing with the impressions of the senses; he does insist, however, that as old age advances the capacity of the mind to draw clusters of sensory ideas together diminishes. Now this argument depended on the substitution of a vibratory model of mind, based on physical properties, for the idea of mind as the essence of a supernatural entity. Mind was no longer conceived by Hartley as the property of a higher being but as a mental state caused by its physical environment, and accordingly the mind must be associated with the physical and material conditions from which it originated; that is to say, the mind was the product of the organism and its relations with the world, and so states of mind were indivisible from states of nature. Through the association of ideas, Hartley argued, we can comprehend the fundamental laws of mental activity; the two principles of vibration (the transmission of impressions) and association (the combination of impressions) underwrote

Hartley's model of mind, and two metaphors, sensation and idea, described its activities.[31] He maintained, in fact, that because the mental and physical realms were mutually determined, an explanation of the law-like relations between sensations and ideas could stand for a description of the conditions under which an individual acts. This sense of parallel between an internal and an external system, both physical nature, where the former is a microcosm of the latter, made the behaviour of individuals in society explicable through their responses to a range of stimuli. More than that, the workings of the internal component of Hartley's system allowed for mind to change physically subject to repeated vibrations in the nerves and the mind.

The importance of this notion of change cannot be overestimated, as it provided physical mechanisms for transformation in respect of human (and animal) action but also deemed instinct or involuntary action as an innate factor in directing actions. Such ideas, particularly with reference to Hartley, are often held to be the basis for the modern tradition of human psychology; in fact, it is clear that they were influential in shaping the theories of change, improvement, and evolution which started to emerge towards the end of the eighteenth century. 'Transformists of the late eighteenth and early nineteenth centuries – particularly Erasmus Darwin, Cabanis, and Lamarck', Robert Richards has said,

exhibited in their theories the force of the sensationalists' discussions of instinct. Under this influence they acknowledged and indeed insisted on the role of intelligence in guiding animal actions . . . The transformists had to show, not only that behaviour and structures were adapted to the environment, but that they were *adaptable*, while yet admitting that behaviour had innate components.[32]

A section of Erasmus Darwin's *Zoonomia*, on 'Generation', amplifies this directly in terms of Hartley's conception of man:

The ingenious Dr. Hartley in his work on man, and some other philosophers, have been of the opinion, that our immortal part acquires during this life certain habits of action or of sentiment, which become for ever indissoluble, continuing after death in a future state of existence; and add, that if these habits are of the malevolent kind, they must render the possessor miserable even in heaven. I would apply this ingenious idea to the generation or production of the embryon, or new animal, which partakes so much of the form and propensities of the parent . . . At the earliest period of its existence the embryon, as secreted from the blood of the male would seem to consist of a living filament with certain capabilities of irritation, sensation, volition, and association; and also with some acquired habits or propensities peculiar to the parent: the former of these are in common with other animals; the latter seem to distinguish or produce the kind of animal, whether man or quadruped, with the familiarity of feature or form of the parent.[33]

The sophistication of Hartley's ideas in positing an integrated model of the mechanisms of mind and body ensured their place in the prehistory of evolutionary ideas about action and behaviour. Yet his conception of man was managed through a fine balancing act between the claims of the physical properties of mind (monism) and a belief in the immateriality of the soul (dualism). What seems to have ensured the balance in Hartley's theoretical system was his use of the moral sense as the pivot between monism and dualism. The term 'moral sense' was notoriously difficult to define but in the broadest terms it was taken to mean an instinctive action towards pleasure or pain and as such offered a prospectus on human mental and moral character. Hartley described its uses and resonances as follows:

The moral sense or judgement here spoken of is often considered as an instinct, sometimes as determinations of the mind, grounded on the eternal reasons and relations of things. Those who maintain either of these opinions may, perhaps, explain them so as to be consistent with the foregoing analysis of the moral sense from association. But if by instinct be meant a disposition communicated to the brain, and in consequence of this, to the mind, or to the mind alone, so as to be quite independent of association; and by a moral instinct, such a disposition producing in us moral judgements concerning affections and actions; it will be necessary, in order to support the opinion of a moral instinct, to produce instances, where moral judgements arise in us independently of prior associations determining thereto.[34]

The connection of instinct, reason, and morality was immensely significant as it enabled actions, affections, dispositions, and judgements to be strung together as the vital constituents of human nature and the determinants of what it is to be human. I suggest that Lavater drew upon Le Brun's theory of the passions or emotions and upon some elements of Hartley's theory of the associative process and employed them as the basis for a descriptive account of man predicated on an instinctive understanding of the purposes and properties of things. What, for instance, is man? What are the essential properties of man? What kind of thing is man? Lavater,

Plate 4   Johann Caspar Lavater, Ten faces of both men and women, 1789.

sometimes directly and often indirectly, suggested answers to these questions.

IV

Johann Caspar Lavater's *Essays on Physiognomy* is a compendious collection of observations and aphorisms which ostensibly contribute to our understanding of man and mind through an eclectic array of illustrations, silhouettes, and descriptions of an individual's character (plate 4).[35] Lavater believed that a description of human nature involved an explanation of the properties or essence of mind and character, and so his account of the nature of man provided patterns for understanding the unity and order of the physical world based on the activity of the mind. Man's essence could be known as long as his actions, gestures, and expressions could be observed, because the state of an individual's mind (and soul) could be derived from these observations; in fact, the teleology to which he subscribed went like this: expression was an index of mind which was, in turn, the spiritual core of man and as such the determinant of an individual's character. The appeal of essentialism for Lavater lay in its capacity to validate what he called a science of mind (or what perhaps we might call a human science) based on a theory of natural kinds or types, but the problem of essentialism for the practice of physiognomy was that it imagined its science as the result of an idealistic understanding of the intrinsic properties and purposes of things. Thus, whilst essentialism underwrote Lavater's science of mind it was also, and not incidentally, the cause of its many inconsistencies.

Fundamental to Lavater's conception of man was an essentialist explanation of character which was based on what was apparent (rather than occult) in human nature. 'Of all earthly creatures, man is the most perfect', he claimed,

the most imbued with the principles of life. Each particle of matter is an immensity, each leaf a world, each insect an inexplicable compendium. Who, then, shall enumerate the gradations between insect and man? In him the powers of nature are united. He is the essence of creation. The son of earth, he is the earth's lord; the summary and central point of all existence, of all powers, of all life, on that earth which he inhabits. (I, p. 10)[36]

Here, Lavater makes manifest the principles of his physiognomic practice: in the first place, man is a unique creature, rendered so as a result of the power and benevolence of the creator; and in the second place, each individual possesses a unity and coherence which marks it out from other human beings. There was no way, Lavater confessed, he could teach mankind the whole of the divine alphabet necessary to translate the language of nature, but he could make some of its characters transparent to the enlightened observer, as the main point of physiognomy was to reveal things in nature, that eluded the immediate comprehension of the senses. Its purpose was, in fact, to disclose

the exterior or superficies, of man, in motion or at rest, whether viewed in the original or by portrait. Physiognomy is the science of knowledge of the correspondence between the external and internal man, the visible super-ficies and the invisible contents. (I, p. 19)

The point is that through physiognomy, the method of reading the external appearance as a sign of the internal state, one could arrive at a definition of man which mapped an individual's inner soul or being onto their external appearance. It might well be that man was flawed, but according to Lavater, physiognomy could instruct and improve man's knowledge of himself, his fellow men, and the Creator of men, by revealing what happened in the mind and ultimately the soul.

At the heart of Lavater's physiognomical system was a description of the natural kinds (or types) of existence which were inherent in the organic world:

To know – to desire – to act – Or accurately to observe and meditate – To perceive and to wish – To possess the power of motion and of resistance – These combined, constitute man an animal, intellectual, and moral being. (I, pp. 10–11)

Though these three kinds could be taken as illustrations of different types of human beings, each had a distinctive character which in theory at least was applicable to animals as well. The first is animal life, localised in the belly and including the organs of reproduction; the second is moral life, focussed on the breast and the heart; and the third is intellectual life, located in the head, with the eye as its central focus. The face was exemplary of these three classes of life, Lavater claimed, in so far as the countenance crystallised the nature of an individual's character. Hence, the mouth and the chin related to animal life; the nose and cheeks represented moral life; and the forehead and eyebrows epitomised intellectual life. Given that physiognomy is the visible expression of certain invisible internal qualities, the idea is that these classifications mark out a hierarchy of description whereby animality is linked above all to the function and structure of the whole human body and provides the lowest order of description, morality is found in the motions of the heart and is the middle order of description, and the intellect corresponds to the head and is the highest order of description (plate 5). To discuss highness and lowness in this form, effectively as localised physical attributes, is significant not least because it suggests a progressivist account of the development of creatures through a series of grada- tions which make explicable, Lavater implies, the distinctions 'from the insect to man'. Thus whilst little attention is paid to the place of animals within this hierarchical system, the acknowledgement of the animal instinct inherent in some types of human beings suggests the possibility of a developmental model of physical change. However, what marks out man from insects, as a distinctive and unique creature is the capacity of the human mind 'to know, to desire, to act'. For Lavater, as we saw in the introduction, the character of human action underpins our relations with other human beings in that we instinctively make quite profound judgements of what we see without considering the reasons for doing so. Physiognomy offered a means of defining and explaining the scope of these instinctive

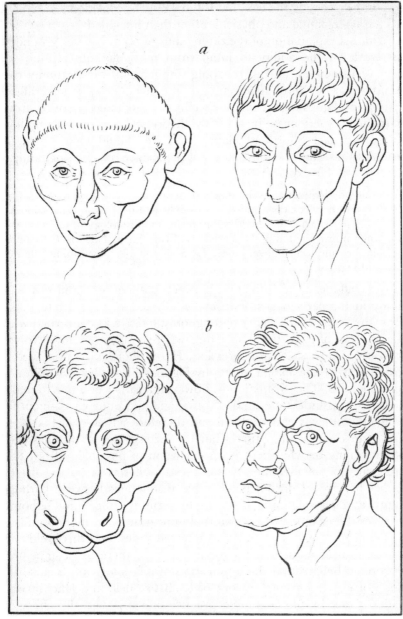

Plate 5   Johann Caspar Lavater, Four heads; comparison of man with monkey and man with bull, 1789.

responses, stressing the fact that we make such judgements in respect of character, actual and physical rather than imagined.

With his explanation of the three kinds of man, Lavater grounded his work on the ability to jump from particular observations to general principles, and in so doing implied that physiognomy can stand squarely within the remit of the sciences as a generalised system of knowledge with a specific methodology derived from particular experiences. In a highly theoretical account of physiognomy, Michael Shortland claims that we should view Lavater's teachings as 'the establishment of a science of physiognomical perception'. He explains:

Scientific physiognomy has the object of arranging, specifying and defining those stable features [as opposed to mobile features] while philosophical physiognomy is what we would today term the physiology of expression – the domain investigated by Descartes, Le Brun, James Parsons, Cureau de la Chambre and John Bulwer before Lavater, and following him by Charles Bell and Darwin.[37]

Lavater alludes to a distinction between pathognomy and physiognomy in the early pages of *Essays on Physiognomy* in order to make clear the scope of his study; pathognomy, he says, is concerned with the analysis of fixed facial structure whereas physiognomy involves the observation of facial expression. Here, Shortland amplifies this distinction to the extent that physiognomy is either scientific or philosophical. The distinction is, however, a rather odd one to make because it assumes one kind of physiognomic practice is scientific whilst the other is philosophical and therefore physiological. It does not seem clear to me either what the criteria are for a science of physiognomy or why there is an exclusion of physiology from this practice; nor, incidentally, is this thesis well served by the list of names he gives. I do not believe that distinctions between pathognomy and physiognomy hold up to scrutiny, since physiognomy involves the examination of external appearances (in motion and in rest) and so provides an index to internal states, and this builds on the same principles as pathognomy. Indeed, there are compelling reasons to believe that the application of physiological ideas to the study of expression in the work of Le Brun, Bell, and Darwin was underwritten by physiognomic practice and also had the effect of annulling its claims about human nature and character.

To use the term 'character' as a means of describing an individual was not, however, unproblematic. A character is basically the property of an organism which involves different states and so produces variations, of the kind usually associated with Darwinian thought. As Lindley Darden has suggested, 'the move to the view that organisms have individual characters, whose *variations* within a population are important objects of study, was indeed an important shift in nineteenth-century biology'.[38] Thus, to ask what kind of thing an organism is involves an examination of the organism in terms of its properties; one might ask what the characteristics or the defining qualities of the creature are, or what makes it what it is. If the object of enquiry is man, and we want to discover what kind of thing man is, then we have to delineate the essential properties of man and demonstrate why, in order to fulfil his function, man must have these particular properties. This essentialist view of an organism, which was at the centre of Lavater's physiognomy, imputes purposes to all things and assigns things to classes on the basis of their possession of common essences. Discrimination, or seeing difference, was physiognomy's approximation to an essentialist method, as it provided a means of reducing the many differences between individual appearances to a few generalised distinctions. To fix difference was therefore, in this case, to see order amidst confusion and glean some kind of human understanding from that which appears (at first sight) to be commonplace – the ordinary occurrences of everyday life. In sum, discrimination is simply another word for essentialism: it functions through a unique taxonomic system which assigns things to classes on the basis of their shared essences.

Discrimination was at the centre of physiognomic practice because in order to discover the universal truths which inhered within specific experiences, the physiognomist must work inductively and judgementally, using particular observations to stand for general types or kinds.[39] All human beings, Lavater proclaimed, have the capacity to make judgements about other individuals on the basis of their difference, and the feeling which underpins this was profoundly physiognomic:

By physiognomical sensation, I here understand 'those feelings which are produced at beholding certain countenances, and the conjectures concerning the qualities of mind, which are produced by the state of such countenances, or of their portraits drawn or painted'. The sensation is universal; that is to say, as certainly as eyes are in any man, or any animal, so certainly are they accompanied by physiognomical sensations. Different sensations are produced in each, by the different forms that present themselves. (I, p. 56)

The physiognomist was privileged in this respect as they could identify the purpose to which such sense or feeling should be put. Though anyone could arrive at a definition of man by inferring a certain kind of character from the appearance of an individual, the idea was that adherence to the teachings of physiognomy enabled the impressions of sense to be translated within a metaphysical system, into higher and lower kinds of individuals. Lavater makes this clear:

Those who have this sense, this feeling, call it which you please, will attribute that only and nothing more to each countenance which it is capable of receiving. They will consider each according to its kind, and will as little seek to add a heterogeneous character as a heterogeneous nose to the face. Such will only unfold what nature is capable of receiving, and take away that with which nature would not be encumbered. (II, p. 101)

To observe physiognomically was, therefore, to contemplate nature and see in its forms evidence of a higher being, something which all human beings could do, and as Christopher Rivers has suggested, 'not only is the sentiment or intuition felt by the physiognomist the first step in the process of interpretation, it is also the most important – indeed, sometimes the only step'.[40]

For Lavater, the first task for the physiognomist is to render each appearance characterful (plate 6); that is to say, to take the external form of an individual as typical of their internal character:

There is no object in nature the properties and powers of which can be manifest to us in any other manner than by such external appearances as affect the senses. By these all beings are characterised. They are the foundations of all human knowledge. Man must wander in the darkest ignorance, equally with respect to himself and the objects that surround him, did he not become acquainted with their properties and powers by the aid of externals; and had not each object a character peculiar to its nature and essence, which acquaints us with what it is, and enables us to distinguish it from what it is not. (I, pp. 11–12)

It was, however, not sufficient to look at a fellow human being in order to comprehend the 'character peculiar to its nature and essence'. Observation was a wide-ranging term, evocative of many meanings, and posed a considerable challenge to Lavater's system. This is what he had to say on the matter:

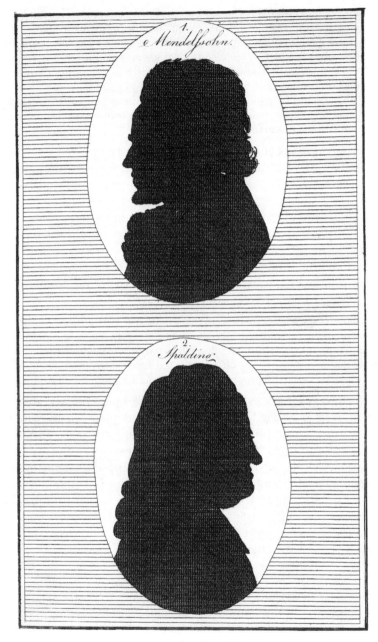

Plate 6   Johann Caspar Lavater, Silhouettes of Mendellsohn and Spalding, 1789.

Precision in observation is the very soul of physiognomy. The physiognomist must possess a most delicate, swift, certain, most extensive spirit of
observation. To observe is to be attentive, so as to fix the mind on a
particular object, which it selects, or may select, for consideration, from a
number of surrounding objects. To be attentive is to consider some one
particular object, exclusively of all others, and to analyse, consequently, to
distinguish, its peculiarities. To observe, is to be attentive, to distinguish
what is familiar, what dissimilar, to discover proportion and disproportion,
is the office of the understanding. (1, p. 119)

The second task for the physiognomist, as indicated here, is to
discriminate the familiar from the unfamiliar, for even though each
observation naturally involved comparison and classification, the
physiognomist was equipped with a special talent for seeing the
general in the particular and the universal in the general. More than
that, the physiognomist (like the taxonomist) sought to pick out
specific from accidental characters and in so doing to segregate
original and habitual expressions from accidental ones. 'There is no
state of mind which is expressed by a single part of the countenance,
exclusively', Lavater wrote; rather,

The whole countenance, when impassioned, is a harmonized, combined,
expression of the present state of mind. Consequently, frequent repetitions of
the same state of mind, impress upon every part of the countenance, durable
traits of deformity, or beauty. Often repeated states of the mind give hability.
Habits are derived from propensities, and generate passions. (1, p. 182)

The purpose of Lavater's teachings was to epitomise the different
kind of responses made by individuals and place them within a
classificatory framework. By rendering the variety of individual
expressions comprehensible in relation to a fixed series of types, the
relation of particular to general observations becomes dependant on
the reduction of complex (individual) expressions to simple (typical)
kinds (plate 7). Indeed, it seems that Lavater is saying that precisely
because we see differently as individuals, there is a need for some
standards of seeing; and it is these standards or norms which, of
course, his physiognomical teachings provided.

v

The practice of physiognomy was designed to discover certain
fundamental principles of human nature through what seems an
idealised form of observation. A science of mind conceived according to physiognomic principles presents an account of the internal

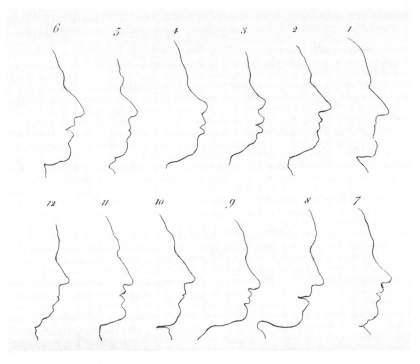

Plate 7   Johann Caspar Lavater, Twelve outlines of idiots, 1789.

processes of the mind as a series of largely instinctive responses to human action which stand for characterisations of the nature of man, and any explanation of how the mind works must satisfy these characterisations; yet they are simply descriptions that detail actual appearances rather than internal processes (such as those which make emotion, instinct, and sensation explicable). What we get, therefore, from a science of mind based on physiognomic principles is not a consistent, systematic, or substantial description of the properties of man or an explanation of the essences proper to man. Rather, we are offered an account of the internal processes of the mind based on a series of instinctive characterisations about the nature of man. To be sure, physiognomic practice held to a democratisation of observation, meaning that it took the fact that we see and judge as explicable in terms of actions and decisions which are universal to man, and it did so through an inductive method which made it impossible for the standards of expression to be much more than subjective measures of qualitative states. In theory, then, Lavater presented an understanding of expression which reports how individuals interact, responding to each other in a specific environment, but in practice he is simply reporting how a system, in this case the mind, is connected to its environment, in this case subjective human relations, *without* offering an explanation of the inner workings of the system.

A scientific or, perhaps more accurately, a naturalistic explanation of man (and the expression of emotion) sits uneasily with the idea of mind as the essence of a higher being: if the mind is the essence of a *super*natural entity then it can only be known by seeing its activities, at second hand as it were, in some of the creatures of the world. Nonetheless, Lavater's physiognomic science of mind seems to depend on precisely these principles; that is to say, it hinges on a willingness to believe not only that there is a world outside the physical world but that this supernatural world can be described in physiognomic terms. This notion of mind as the spiritual essence of a higher order is heavily reliant on some sort of intuition of the properties and natural purposes of mind, and this, for Lavater, was one of its greatest strengths. So, for instance, an anonymous essay in *The Gentleman's Magazine* (1808) suggested that written references could be replaced by a physiognomic mode of interpreting the face of prospective employees: 'The giving of characters, a duty which is seldom faithfully performed, either from pique or want of discernment, might have been dispensed with, each applicant carrying a certificate in his, or her forehead, eyes, nose, or chin, and would have been readily supplied with places, according as their masters had a confidence in one feature more than another.'[41] However ludicrous this idea of certificating the face now appears, it was on these grounds that physiognomy and its practices experienced a marked revival. Over seventy years after its original publication, in fact, the eighth edition of the *Encyclopaedia Britannica* (1853–60) described Lavater's work in wonderfully hyperbolic terms: 'its publication created everywhere a profound sensation; admiration, resentment, and fear were cherished towards the author. The discoverer of the new science was everywhere flattered or pilloried; and in many places, where the study of human character from the face became an epidemic, the people went masked through the streets.'[42]

In the next chapter we will see how physiognomy and its

characterisations of human nature become more systematic, and indeed scientific, in the hands of Charles Bell. For Bell, the interest of physiognomy (in the form developed by Lavater) lies in what it can tell us about the relationship between expression and emotion, and whilst physiognomy verifies a natural theological approach to the physical world, it is clear that it also invites reflection on the nature of both instinct and observation. Natural theology employs notions of contrivance and consciousness through the concept of design but, as we will see, its philosophical underpinnings are directly related to Lavater's and Bell's attempts to explain the structure and function of the organic world through an examination of the actual appearances of its creatures, and in particular man. Bell's work is significant because it appropriates new physiological explanations of structure and function, within the context of natural theological principles, to explain our understanding of the fit between expressions and their emotional meanings.

CHAPTER 2

The argument for expression: Charles Bell and the concept of design

The operations of the mind are confined not by the limited nature of things created, but by the limited number of our organs of sense.

Charles Bell[1]

I

In a letter to his brother, written in August 1830, Charles Bell exclaimed: 'Behold, with what I point . . . This hand, how exquisite in form and motion. But first turn over – use it and learn to admire!' The letter continued, uncharacteristically displaying the exuberance and jocularity of the author:

I have a letter this morning from the President of the Royal Society, who, with the counsel and approbation of the Archbishop of Canterbury have proposed to me to write on the Human Hand . . . I think I know now what to engrave on my seal – a hand. I shall introduce it on all occasions, sometimes doubled . . . as implying the pugnacious nature of the man – sometimes smooth and open, as ready to receive – sometimes pointing, as from the master. In short, I shall make use of this hand until they acknowledge me a handy fellow![2]

His pleasure in the scientific and religious commendation implicit in the invitation is evident in the rhetorical play upon the verbal and visual associations of the word 'hand'. Thus the 'doubled' hand represents fighting spirit whilst the 'pointing' hand refers beyond man to the actions of a creator. The impetus for the letter was an invitation to write a Bridgewater treatise, a series of books by leading scholars of the times which were intended to extol certain scientific theories within the context and ideological leanings of natural theology. With a dedication to 'the Power, Wisdom, and Goodness of God, as manifested in the Creation; illustrating such work by all reasonable arguments',[3] and the subject matter predetermined in

44

the will of the third Earl of Bridgewater, the treatises made no secret of their religious affiliation and instead offered an idealistic account of subjects such as anatomy, physiology, and astronomy.[4]

There were nine official treatises in all – written between 1833 and 1837, and including Peter Mark Roget's *Animal and Vegetable Physiology* (1834) and William Whewell's *Astronomy and General Physics* (1833) – which shared the professed intention to address a general middle-class audience seeking material proof of a divine presence in the natural world. Throughout, the authors sought to negotiate the distinctive claims of both religious and secular philosophies, and as a result statements of the following kind, in this case by Bell, were common:

Nothing affords a more perfect proof of power and design, than the confidence all men put in the correspondence between the perceptions or ideas that arise in the mind, through the exercise of the organs of the senses, and the qualities of external matter. Although it must ever be beyond our comprehension, how the object presented to the outward sense and the idea of it are connected, they are, nevertheless, indissolubly united; so that the knowledge of the object, gained by these unknown means, is attended with an absolute conviction of the real existence of the object – a conviction independent of reason, and to be regarded as the first law of our nature.[5]

This is a conceptually complex passage for it alludes to the actions of a higher being with the term 'design', but also stresses the physical organisation of man and in particular, the receptions of impressions in the mind. It is, in fact, a thoroughly transcendental view of the world which links human action to physical phenomena in so far as they are both explicable as products of the mind, even if the mental processes remain nevertheless unaccountably mysterious. Actually, though the concept of design had to be taken as the starting-point for the treatises, the work of Whewell, Roget, and Bell manifests a developing tension between a theological investment in the theory of mechanical organisation and the materialist idea that natural phenomena had a coherence of their own.

This chapter examines the tensions inherent in this discussion, exploring the arguments of Bell and his contemporaries for the specific kind of conceptual framework adequate to organise the organic world and explain the conditions of human existence. I am not claiming for Bell an especially prominent or radical position in the history of nineteenth-century scientific thought; rather, I am claiming that his work is important in developing physiological explanations of mental processes, and especially emotional expression. In an important essay on Bell, Ludmilla Jordanova frames Bell within the scientific culture of the early nineteenth century and views his writings as a product of that specific historical context. 'It is important to search for the underlying unity', she writes, 'not just in Bell's preoccupations but in his contacts with others':

His form of conservatism was quite historically specific. It seems that neither art historians nor medical historians have known quite what to do with Charles Bell, or, indeed with other figures like him. In proposing ways of examining him, I am therefore speaking about Bell both as an unusually rich case study for social and cultural history and as an example for historiographical debate.[6]

Born in Edinburgh in 1774, Bell received his early training in the 'Arts of Surgery & Pharmacy' as an apprentice to his elder brother John, Professor of Medicine at the Edinburgh School of Medicine.[7] Whilst still a medical student in Edinburgh, Bell published *A System of Dissections, Explaining the Anatomy of the Human Body, the Manner of Displaying the Parts, and their Varieties in Disease* (1798), a work of consummate detail which displayed a knowledge of anatomy impressive enough to secure Bell's position alongside his brother John as an expert in anatomy and surgery.[8] Strongly influenced by the Scottish Enlightenment tradition, Bell and his brothers were part of a large circle of Edinburgh gentlemen, old-school Whigs, including Francis Jeffrey, Sydney Smith, and Henry Brougham, founders of *The Edinburgh Review* (1802), who held to a conviction that the phenomena of the physical world were ultimately reducible to exact description and explanation.[9] It was a conviction Bell and his friends shared with Lavater, whose legacy to Bell was a view of the organic world ordered according to a series of natural laws that underpinned the infinite variety of forms accessible to the casual observer.

By postulating a universe within which the physical laws ordained by its creator were not self-evident, the main function and purpose of Lavater's physiognomy was to identify and describe the common forms which organised the diversity of appearances. To Lavater, these forms or appearances functioned in a profoundly normative manner as determining what was common to all people and all things in the physical world. Thus, physiognomy explained human nature in terms of a uniform order of types or kinds and it did so at the level of the individual, translating particular observations into general (and quite often imaginary) interpretations of character and emotion; as such, its effects were as much metaphysical and moral as they were ideological. The challenge for Bell, therefore, was to identify the grounds upon which his own physiognomical teachings could be established as true without denying the validity of his physiological ideas. Were physiognomy and its teachings essentially a product of empirical (scientific or artistic) observation? Or, were they conclusions derived from an anatomical understanding of nerves and muscles? Was the artist obliged to begin from a thorough grounding in the knowledge of anatomy? Must the artist thus be obedient to science? My purpose in this chapter is to consider the ways in which Bell attempts to answer these questions, particularly in respect of the relationship between the theological conception of creation and the physiological notion of sensation. I shall start by looking at the concept of design, outlining the claims of its principal proponent, William Paley, for the importance of contrivance and consciousness to the natural theological explanation of the world and then contrasting them with the comparative anatomy of Georges Cuvier, both of whose work was familiar to Bell. I shall then evaluate Bell's use of design in two works, a short and significant essay on the 'Idea of a New Anatomy of the Brain'(1811) and his Bridgewater Treatise on *The Hand* (1833),[10] before turning to his argument for expression, so-called, in *The Anatomy and Philosophy of Expression, as connected with the Fine Arts* (1844).[11] The value and significance of Bell's work is, I suggest, twofold: in the first place, it affirms the importance of the divine aesthetics of the physiognomical tradition for the study of expression; and in the second, it signals the import of the new physiological doctrines in providing inner, scientific rationales for the expression of emotions.

II

A conception of the place of man and mind in nature was crucial to debates about the relation of life and matter in the first third of the nineteenth century. As the epigraph to this chapter reveals, Bell's conception of nature suggested to him that it was not the 'nature of things created' but the 'number of our organs of sense' which limited the activity of the mind. This is an important distinction, as Bell proposes that our organs of sense mediate between the mind and the external world and in so doing are dependent on the laws of the physical world; thus the occurrence of sensation itself becomes a deciding factor in colouring our understanding of the nature of human existence. Departing somewhat from Hartley, who emphasised every stage of the activity of mind including the transmission of impression and the resulting ideas (of sensation and intellect), here Bell suggests there are physical constraints on our mental activity. Implied in his statement is an allusion to the creative acts of a higher being who, it would seem, controls the responses of individuals to their environment. The underlying issue was the extent to which an individual had control over their life as constituted by body and mind, and, as L. S. Jacyna has pointed out, 'it was difficult, if not impossible, to talk of life without at once impinging on such topics of the greatest sensitivity as the nature of the soul and the relation between God and man'.[12]

Loss of control was a topic of some concern at the time, whatever form it took, because it posed a considerable challenge to descriptions of the order and regulation of society, and yet enabled order to be defined against disorder, and control to be contrasted with a lack of inhibition. In an intelligent study of the concept of inhibition, Roger Smith claims that 'emotional outbursts, childish behaviour, drunkenness, dreams – such common experiences threw into relief the ideal of a rational, conscious, and well-regulated life'.[13] The word 'inhibition' and its conceptual fashionings are important, Smith contends, because it emerges in common usage at the end of the nineteenth century, the product of more than a hundred years' thought on human nature and the nature of the human mind. Hartley's associationist psychology is, of course, part of this tradition in that one of its principles was the correlation between the mind and body so that 'order in mental content reflected the order (contiguity and similarity) of sensations in experience'.[14] Indeed, Smith cites the opinion of the influential physician Henry Holland, who stresses the importance of looking at the conditions under which control is lost as a means towards greater understanding of the mind:

Dreaming – insanity in its many forms – intoxication from wine or narcotics – and the phenomena arising from cerebral disease, are the four great mines of mental discovery still open to us; – if indeed any thing of the nature of discovery remains, on a subject which has occupied and exhausted the labours of thinking men in all ages . . . By the curtailment of suspension of certain functions, by the excess of others, and by the altered balance and connection of all, a sort of analysis is obtained of the nature of mind.[15]

The whole idea of balance, in life as in mind, was extremely significant to Bell; not only does it provide a common thread throughout his diverse writings but, more importantly, it also suggests the importance of natural theology as the means of explaining the order and regulation of the organic world.

Natural theology presented a supremely universalist view of the world which was based on an understanding of the correspondences between the essences of all things in the world and the power and intelligence of a higher being. A harmonious picture of the organic world was conceived which depended on a number of laws derived from the processes of natural phenomena. Revelation was the key to understanding these laws as it enabled the divine plan to be demonstrated, even proved, in terms of these natural processes. It was a compelling vision of the order and control inherent in the world whereby natural processes were equated with revealed processes. One of its earliest and most explicit statements can be found in Bishop Joseph Butler's notable work, *Analogy of Religion* (1736). He wrote:

We know indeed several of the general laws of matter: and a great part of the natural behaviour of living agents is reducible to general laws. But we know in a manner nothing, by what laws, storms, and tempests, earthquakes, famine, pestilence, become the instruments of destruction to mankind. And the laws, by which persons born into the world at such a time and place are of such capacities, geniuses, tempers; the laws, by which thoughts come into our mind, in a multitude of cases; and by which innumerable things happen, of the greatest influence upon the affairs and state of the world; these laws are so wholly unknown to us, that we call the events which come to pass by them, accidental: though all reasonable men know certainly, that there cannot, in reality, be any such thing as chance, and conclude, that the things which have this appearance are the result of general laws, and may be reduced into them.[16]

The subject here is the extent to which the actual appearances of things can be ascribed to invisible and often unknown causes. How, in other words, do we account for the appearances of things in nature – do we look to their behaviour, their constitution, or their development? One answer, as Gillian Beer has argued with a nod to Butler, can be found in the use of analogy and its mystical, almost magical effects. Analogy, she explains, 'claims a special virtue at once incandescent and homely for its achieved congruities. A *living*, not simply an imputed, relation between unlikes is claimed by such discourse'.[17] Beer's suggestion is that analogy supports a singular order of things but by aligning comparable elements is able to reveal the mechanisms which lie behind (and explain) the natural processes of things. The problem is the laws. Under what conditions might the structures of things be revealed through their function? Also, to what effect? Or, to put it slightly differently, can revelation satisfy the desire, in Bell at least, for teleology?

Derived mainly from the teachings of William Paley, natural theology posited the interrelation of all phenomena and in so doing, used terms like design, mechanism, order, and unity to convey the sense of a Creator pursuing a grand plan and instilling purpose in all living things.[18] Paley's classic statement of these ideas in his *Natural Theology* (1802) was as follows:

In crossing a heath, suppose I pitched my foot against a *stone*, and were I asked how the stone came to be there, I might possibly answer that, for anything I knew to the contrary, it had lain there for ever; nor would it, perhaps, be very easy to show the absurdity of this answer. But suppose I had found a *watch* upon the ground, and it should be inquired how the watch happened to be in that place, I should hardly think of the answer which I had before given – that for anything I knew, the watch might have always been there. Yet why should not this answer serve for the watch as well as for the stone? Why is it not as admissible in the second case as in the first? For this reason, and for no other – namely, that when we come to inspect the watch, we perceive (what we could not discover in the stone) that its several parts are framed and put together for a purpose.[19]

The logic of this passage, from appearance and chance to mechanism and purpose, exemplifies the Paleyite interpretation of nature which explained the relation of both stone and watch to their natural environment in terms of purpose or, more accurately, purposiveness. Paley proposed that the place and subsequent adaptation of any thing in its environment was a sign of the intelligence of its design and the benevolence of its designer, both of which help us to infer the presence of a higher being who guarantees that the world we live in is the best possible world in which we could live. Thus,

there cannot be a design without a designer; contrivance without a contriver; order without choice; arrangement, without any thing capable of arranging; subserviency and relation to a purpose, without that which could intend a purpose; means suitable for an end, without the end ever having been contemplated, or the means accommodated to it. Arrangement, disposition of parts, subserviency of means to an end, relation of instruments to an use, imply the presence of intelligence and mind'.[20]

If we put this together with the previous quotation, a clearer picture of the organic world, its organisation and different mechanisms, starts to emerge. Nature had its analogue in the form of a book in so far as both must be read carefully and closely in order to comprehend the hidden meanings which lie behind their material forms. The practice of reading which natural theology relies on is, in fact, a singular one, for the meanings contained in nature's forms are not unlimited but delimited by the benevolent control of the creator. Thus, the design of the world was planned and motivated through the regulation of a higher being, and as Charles Gillispie has suggested, 'the way to find out the will of God is to find out what works; since it works, God must have intended it to work, and it is, therefore, His will'.[21] The point is that if we are to find out 'what works' then we must labour at our task; whilst Paley stressed the clockwork movements of the natural order, he admitted there is nothing easy or straightforward in seeing the design which lies behind the appearances of things – he believed the intellectual effort necessary to see God served only to increase our appreciation of the contrived, creative, and personal attributes of the divine mind.

With natural theology we have, then, a system which was designed to prove the existence of a divine being, at once mind and maker, by attending to the way things work. Design was akin to a morphological system of understanding – meaning that it focussed on the physical form of things – that outlined the scope and rationale of what could be known on an extra-phenomenal level, and so was deemed what can quite literally be seen as the vehicle for higher and more sophisticated forms of knowledge. These are Paley's thoughts on the matter:

The existence and character of the Deity is, in every view, the most interesting of all human speculations. In none, however, is it more so, than as it facilitates the belief of the fundamental articles of *Revelation*. It is a step to have it proved, that there must be something in the world more than what we see. It is a further step to know, that, amongst the invisible things of nature, there must be an intelligent mind, concerned in its production, order, and support. These points being assured to us by Natural Theology, we may well leave to Revelation the disclosure of many particulars.[22]

We cannot, Paley argued, have a watch without a watchmaker, but, he cautioned, we should not expect to know the watch or its maker

fully or completely in our present state because perfect happiness
can only exist in a future state. So, just as the notion of mechanism
determined the way things worked in a Paleyite world, so the
category of nature, with its processes and laws partially played out in
this world, proved the existence of a divine mind. The adoption of
the transcendentalist thesis integral to natural theology sustained a
belief in the separate existence of the soul and as such perpetuated a
dualist vision of the distinction between body (life) and mind (spirit)
for individuals and their social relations. 'Moral order was impossible to maintain', Jacyna says, 'unless the soul were immortal and
the Deity equipped with intelligence and moral attributes . . . Just as
matter was brute and inert unless pervaded and controlled by the
emanations of God's will, so men were incapable of a civilized
society without divine government.'[23]

There are many ramifications of these ideas, but for Bell they
provided a means of conceptualising an organic world, controlled
and regulated in physical, mental, and social terms, and as such
holding the line of his conservative ideals. The importance of
natural theology for him (and many of his contemporaries) was that
it made theological ideas demonstrable and so in theory compatible
with a more naturalistic view of life, thereby ensuring that an
investment in a theory of mechanical organisation did not constitute
a denial of theological doctrine. Whilst there was no question that
for Bell design provided the conceptual framework which organised
nature and the conditions of human existence, his difficulty was how
to reconcile his practical investment in natural theology with a
growing interest in the possibility that every act of sensation was
itself a living phenomenon and so subject to physical rather than
theological laws. The most sustained exposition of these ideas can be
found in Bell's Bridgewater treatise on *The Hand*, despite the fact that
it was conceived within narrow constraints. As one notable opponent
of the application of natural theology to the practice of medicine
remarked, Bell 'never touches a phalanx and its flexor tendon,
without exclaiming, with uplifted eye, and most reverentially-con-
tracted mouth, "Gintilmin, behold the winderful eevidence of
desin." '[24] Bell's interest in physiological explanations of the relationship between mind and body did not sit easily with his use of design
to explain the functional bases of life, as we shall see shortly, for
though his early writings express the belief that mental activity
cannot be reduced to physiological analysis, his later work struggles

to find a way of placing the interpretation of mind within the context of anatomical and physiological examination. The latter was after all an explanation of the physical world which made function the premise for structure in living creatures; in this it was heavily dependent on the Aristotelian theory of biological form and endorsed by the comparative anatomy of Georges Cuvier in the early years of the nineteenth century.

III

As professor of comparative anatomy at the Muséum d'Histoire Naturelle and permanent secretary of the Académie des Sciences in Paris, there is little doubt of the power wielded by Cuvier in French medical and zoological matters. The proponent of a 'conservative, factual, and safe science', Cuvier's comparative anatomy derived from a zoological concern with both physiological and anatomical questions about the nature and activity of organic life.[25] This meant that the functional properties of all living things could be explained in terms of their chemical, mechanical, and structural processes. With the activity of each thing or organism paramount, Cuvier's functionalism was driven by a desire to explain the use to which an individual part was put within the context of the whole organism. So, for instance, the forelimb (in quadrupeds) is intended to be used for flying, swimming, or walking, and its flexibility, structure, shape, and size are all adapted to serve these ends and fulfil this nature. He explained this as follows:

Natural history nevertheless has a rational principle that is exclusive to it and which it employs with great advantage on many occasions; it is the *conditions of existence*, or popularly, *final causes*. As nothing may exist which does not include the conditions which made its existence possible, the different parts of each creature must be coordinated in such a way as to make possible the whole organism, not only in itself but in its relationship to those which surround it, and the analysis of these conditions often leads to general laws as well founded as those of calculation or experiment.[26]

Cuvier held that there was a natural hierarchy amongst phenomena in the organic world. There were, he argued, four distinct divisions or *embranchements* in the organisation of animal life – vertebrates, molluscs, articulates, and radiates – characterised by their nervous systems and yet sharing structural similarities because of function.

His thesis was that the structure of an animal provided it with the perfect means of functioning in its particular environment not only because it was sufficiently adapted (via its internal organisation) to its habitat but also because its function explained its structure. The link to Aristotle is clear here:

Now, as each of the parts of the body, like every other instrument, is for the sake of some purpose, viz., some action, it is evident that the body as a whole must exist for the sake of some complex action. Just as the saw is there for the sake of sawing and not sawing for the sake of the saw, because sawing is the using of the instrument, so in some way the body exists for the sake of the soul, and the parts of the body for the sake of those functions to which they are naturally adapted.[27]

The nervous systems of each *embranchement* in Cuvier's system were, in fact, quite discrete arrangements, and as a result his vision of nature left large gaps between the divisions of its system. It was a discontinuous system without gradation or transitional development or the possibility of progression or transformation, constructed on the performance of particular roles within a fixed and correlative species: there was no way, following the logic of Cuvier's thought, that nature could be comprehended without God, nor, on the same basis, that an organism could be explained without function. According to Adrian Desmond's important study, 'because function was the sole arbiter of structure, the shape and number of the elements in any organ could be varied by God, or new elements could be added to suit individual needs. Also, because every organ was integrated and functioning perfectly, there were no rudimentary or useless organs, introduced solely to complete some vertebrate "plan".'[28]

It is clear from his writings that Bell was heavily influenced by Paley's natural theological system and also, I think, that he drew upon the functionalism of Cuvier (who he met at least once in London in 1830). This is not to suggest that Paley and Cuvier were at one, intellectually or theologically, though neither were they on opposing sides; rather that Bell takes their thought on board and mediates it in his own writings. The distinctions between Cuvier's functionalism and Paley's design have been well documented, but it is worth summarising a couple of the main points here.[29] The first is that Paley and his natural theologians believed that adaptation was evidence of the intelligence of design, whereas Cuvier and his comparative anatomists held there could be no grand design beyond what was evident from the interrelation of individual organisms. According to Desmond,

Cuvier spearheaded the trend away from older Enlightenment systems of nature toward a new specialization and professionalism, which concerned itself with facts, description, and low-level laws of correlation. This move had its conservative dimension, shown by the fact that Cuvier's empiricism became exaggerated when the political temperature rose in the 1820s and rival deistic theories of form and progress began to be purloined as republican ammunition.[30]

The second, consequently, is that the common plan (or blueprint) operated on a macro-level for Paley and on a micro-level for Cuvier, with corresponding economies of scale. In effect, Cuvier subscribed to a metaphysical dualism which was in tune with natural theology (and Bell's account of expression) but not with his physiological writings which tended towards materialism.[31]

The starting point for Bell's 'Idea of a New Anatomy of the Brain' (1811) was an investigation of sensation and the movements of the nerves. As we saw in the previous chapter, discussions of human nature and the physical actions integral to it tended to involve consideration of the will, along with consciousness and the soul, in determining the voluntary and involuntary movements of the mind and body.[32] The idea of reflex action was all-important in these discussions as it involved giving an important place to the nervous system in any investigation of motion, and often had the corollary that there was a central place – the *sensorium commune* – where sensory nerves meet motor nerves thus enabling sensory impressions to be translated into muscular actions.[33] A theory of sensation was central to an understanding of the nervous system and defined the scope of any investigation into the interrelation of mind and body, in particular the organisation of matter. Bell's role in these debates was significant, for he discovered in 1811 that the anterior root of the spinal chord and the cerebrum had both motor and sensory functions whereas the posterior root directed the involuntary actions of the body. 'The spinal nerves being double, and having their roots in the spinal marrow', he wrote, 'of which a portion comes from the cerebrum and a portion from the cerebellum, they convey the attributes of both grand divisions of the brain to every part'.[34] The important point was that even though Bell identified the motor function of the anterior roots, he paid scant attention to the sensory attributes of the posterior roots.[35] For Bell, the actions of the nerves had not been fully accounted for by existing studies of the operations of the brain, whereas he believed the nervous system involved the activity of mind at every stage of its processes, an opinion which deemed the brain central to physiological explanations of neural function. In a very early and quite cautious stab at theorising these ideas, prior to writing his anatomy of the brain, he said:

There can be no natural division of the nervous system for it is a whole so connected in function, that no one part is capable of receiving or imparting any sensation or of performing the operation of the intellect ... No sensible man will expect, in the most minute and unwearied investigation of the structure of the brain to find the explanation of its function. It is interesting to find effects so peculiarly connected with the operations of the mind, depending upon a structure of so gross and animal nature of the brain.[36]

It was on precisely this subject that Bell constructed an anatomical theory of the brain which, as Karl Figlio has argued persuasively, was much more than an adumbration of his later ideas on the function of the spinal nerve; it was, in fact, 'a critique of a traditional model of neurological activity'.[37]

Viewed as *sensorium commune*, the brain was generally thought to receive impressions from external objects through the nerves and imprint sensations upon a passive mind. Interestingly, Cuvier described the import of this idea in familiar terms:

The nature of the sensitive and intellectual principle is not at all in the realm of natural history; but it is a question of *pure anatomy* to know to what point of the body the physical agents which occasion sensations must arrive, and from what point those which produce voluntary movements must depart, in order that these sensations and movements occur. It is this, the terminus of our passive relations, and the source of our active relations with the external world, that has been named *the seat of the soul, or the sensorium commune.*[38]

There are three important issues here for Bell. The first is the contrast this 'simple, passive impression model of sensation' presents to the associationist psychology of Hartley which we looked at in chapter one.[39] It highly probable that Bell would have been familiar with Hartley's ideas, particularly as it appears that Bell rejected the simple model of the *sensorium commune* (which specified a single point of mental activity) in favour of a complex form of mental activity more like Hartley's model of mind and body. This is his explanation:

In opposition to these opinions, I have to offer reasons for believing, That the cerebrum and cerebellum are different in function as in form; That the parts of the cerebrum have different functions; and that the nerves which we trace in the body are not single nerves possessing various powers, but bundles of different nerves, whose filaments are united for the convenience of distribution, but which are distinct in office, as they are in origin from the brain.

That the external organs of the sense have the matter of the nerves adapted to receive certain impressions, while the corresponding organs of the brain are put in activity by the external excitement: That the idea or perception is according to the part of the brain to which the nerve is attached, and that each organ has a certain limited number of changes to be wrought upon it by the external impression:

That the nerves of sense, the nerves of motion, and the vital nerves, are distinct through their whole course, though they seem sometimes united in one bundle; and that they depend for their attributes on the organs of the brain to which they are severally attached.[40]

The second and related point is that Bell's rejection of the 'simple model' was based on a physiological account of sensation and an anatomical explanation of the brain. In effect, the rigid mind/body distinction of the Cartesian world view was weakening in the face of a growing awareness of the correlation between sensation and impression. Having despatched 'the prevailing doctrine of the anatomical schools'[41] which asserted the unified structure of the brain, Bell identified his model as follows:

When this whole was created . . . the mind was placed in a body not merely suited to its residence, but in circumstances to be moved by the materials around it; and the capacities of the mind, and the powers of the organs, which are as a medium betwixt the mind and the external world, have an original constitution framed in relation to the qualities of things. It is admitted that neither bodies nor the images of bodies enter the brain. It is indeed impossible to believe that colour can be conveyed along a nerve; or the vibration in which we suppose sound to consist can be retained in the brain: but we can conceive, and have reason to believe, that an impression is made upon the organs of the outward senses when we see, or hear, or taste.[42]

The final point (taking us back to the epigraph) is that any imperfections or weaknesses in our vision arise as a result of the limits of the senses rather than the limits of the mind. Hence,

If light, pressure, galvanism, or electricity produce vision, we must conclude that the idea in the mind is the result of an action excited in the eye or in the brain, not of any thing received, though caused by an impression from without. The operations of the mind are confined not by the limited nature of things created, but by the limited number of our organs of sense. By induction we know that things exist which yet are not brought under the operation of the senses. When we have never known the operation of one of the organs of the five senses, we can never know the ideas pertaining to that sense; and what would be the effect on our minds, even constituted as they now are, with a superadded organ of sense, no man can distinctly imagine.[43]

In other words, whilst our organs of sense mediate between the mind and the external world they are, nonetheless, dependent on the laws of the physical world prescribed by God. Because a teleological system like natural theology offered an explanation of present phenomena in terms of future and final causes, a direct connection was assumed between ends and events which made nature appear purposeful.

IV

We have seen how under the direction of Paley the future end of the physical world was always dependent on the creative act of its divine author. Nowhere was this purposefulness more explicitly played out than in the example of the hand. What sequence of events was necessary to produce the hand? What of that apparent anomaly in its structure and function, the thumb? One of the most common arguments went like this: man's unique mental capacities required an instrument like the thumb in order to perform to the best of his abilities and satisfy his desires, therefore man has been endowed with the thumb which is a versatile organ. The abiding fascination of Bell's treatise on the hand lies in his efforts to mediate the concept of design (and its purposiveness) with the wider concerns of his interest in the application of scientific theory to living and particularly human forms. Of course, the consistent use of the hand as an anthropomorphic symbol in literature and history provided Bell with a theological context in which to place his work, but what evolves in *The Hand* is more than a theological narrative of the presence of God's power directing the natural world, for Bell undertook to develop his existing theories of mind and sensation by employing the hand as a vehicle for description and dissection. The hand is a unified but invisible structure, a display of natural design, and a figure of writing: it is, at once, a motif of character, a 'seal', a mode of performance, 'doubled . . . smooth and open', and an indicator of

God, 'pointing'. As a part of the body invested with such an array of potential interpretations and representation, the hand is Bell's central organ of expression and as such is used as an illustration of the structure and function of the body from the inside and the outside. That is to say, the hand acts on different levels: as an exposed surface upon which (metaphorically) to read and to write outside the body; a tactile experience of sensation inside the body; and as a double-edged instrument of instruction to describe and dissect the surface and the sensations of the body.

In a theoretically intricate and technically entangled passage from the last chapter of *The Hand*, Bell gathered together his ideas on the simple, but often overlooked, relation between the hand and the eye in order to illustrate the difficulty of visualising the anatomy of the brain and the nerves of the body. Explaining the importance of conceiving the eye as part of the 'muscular apparatus' of the body – that is, 'its exterior appendages of muscles . . . its humours and the proper nerve of vision' – he directed the reader as follows:

When, instead of looking upon the eye as a mere camera, or show-box, with the picture inverted at the bottom, we determine the value of muscular activity; mark the sensations attending the balancing of the body; that fine property which we possess of adjusting the muscular frame to its various inclinations; how it is acquired in the child; how it is lost in the paralytic and drunkard; how motion and sensation are combined in the exercise of the hand; how the hand, by means of this sensibility, guides the finest instruments: when we consider how the eye and hand correspond; how the motions of the eye, combining with the impression on the retina, becomes the place, form, and distance of objects – the sign in the eye of what is known to the hand: finally, when, by attention to the motions of the eye, we are aware of their extreme minuteness, and how we are sensible to them in the finest degree – the conviction irresistibly follows, that without the power of directing the eye, (a motion holding relation to the action of the whole body) our finest organ of sense, which so largely contributes to the development of the powers of the mind, would lie unexercised.[44]

According to Bell, the value of the eye lies with its capacity for guiding, corresponding, positioning. However, the eye is analogous to the hand only if its activity is experienced as a combination of 'motion and sensation' which involves perception and depends on 'the balancing of the body'. Unlike the immediacy of the sensation experienced by the hand, the activity of the eye is experienced via the translation of muscle into sense, and as such it acts as a summarising sign of what the hand has already felt. The 'motions of the eye' are, Bell argues, 'the sign in the eye of what is known to the hand' namely because touch rather than sight is the determining link between internal and external function in physical life; it is 'the common sensibility. . . the most necessary of the senses . . . enjoyed by all animals from the lowest to the highest in the chain of existence'.[45] Although it was not particularly unusual to place the experience of seeing as a sensation inside the body in the early nineteenth century, Bell's treatise is instructive for the attention it draws to the physiological structure of the body, and in particular the subjective experience of emotion and sensation which can be felt through muscular reflexes.

By aligning the tactile and optical sensorium in a symbolic way, Bell forged an alliance between muscle and sensation that not only privileged the hand as an educational instrument and a source of exquisite sensations – focussing on the eye's function as a muscular activity that occurs inside the body – but also emphasised the resulting phenomenalism founded on muscle sensation. Having discarded the idea of the eye as a camera and collapsed the distinction between internal and external forms, Bell figured the act of looking as an act of touch which makes the individual (physiologically and psychologically) conscious of 'the place, form, and distance of objects'. At stake is 'the power of directing the eye' and, as Gillian Beer has suggested, it is almost as if the movements of the eye involved in this particular act of looking delicately caress the external world with each movement of the body.[46] The reason is simple: 'accompanying the exercise of touch', Bell pointed out in an earlier passage, 'there is a desire of obtaining knowledge; a determination of the will towards the organ of the sense'. And he carried on: 'In the use of the hand there is a double sense exercised; we must not only feel the contact of the object, but we must be sensible to the muscular effort which is made to reach it, or to grasp it in the fingers.'[47] What Bell strove to communicate, then, was the dual function of the hand: it embodies the physical feeling experienced when objects touch, and at the same time transmits impressions of that feeling to the brain. The difficulty was how to visualise this complicated connection between touch and sight and so convey the physical state (composed of muscles and nerves) associated with the hand.

To the artist and the anatomist, the hand operates through the sensibility of touch, is figured by the scalpel and the pen, and in its twin processes of dissection and description reveals its complicity with the eye. The sensibility of touch is aligned with the sensibility of sight as a method of observation which looks both inwards and outwards, first exploring its own body and then making contact with other bodies: 'the knowledge of external bodies as distinguished from ourselves, cannot be acquired until the organs of touch in the hand have become familiar with our own limbs'.[48] Bell draws the observer into the body to feel the effect of the sensations of the muscular frame and also to participate in the production of figures which it observes. The hand and the eye become interchangeable as synaesthetic modes of perception which rely upon the symmetry of the body in order to locate the experience of seeing as sensation occurring inside its structure. The hand is the technical instrument which opens up the structure of the human figure in order to determine the essential characteristics of the figures of expression manifested on the surface of the body. So the action of the hand outside the body is stimulated by the action (expansion and contraction) of the muscle fibres inside the body. As Bell explained, like the face the hand makes manifest through its muscular action the network of relations which join muscles and nerves; it indicates the robust physiological structure of the physical body whilst simultaneously suggesting the delicacy of its neurological impressions:

With the possession of an instrument like the hand there must be a great part of the organisation, which strictly belongs to it, concealed. The hand is not a thing appended, or put on, like an additional movement in a watch; but a thousand intricate relations must be established throughout the body in connection with it – such as nerves of motion and nerves of sensation.[49]

It was just these 'intricate relations' of the nerves of motion and sensation which were explored in detail (with reference to the expression of the emotions) in Bell's study of expression, *The Anatomy and Philosophy of Expression, as connected with the Fine Arts* (1844).

V

The object of Bell's study of expression was to demonstrate the relevance of a knowledge of anatomy to the fine arts, proving the ways in which anatomy could illuminate the truth of expression and of character. Anatomy was 'the examination of that structure by which the mind expresses emotion, and through which the emotions are controlled and modified', and as such was immensely relevant to

painterly practice. Only through a knowledge of the internal structure of the body could the responses of the body to various emotions
not only be accurately observed but also translated into a visual
language. 'The organs of the body', Bell claimed, are 'the links in the
chain of relation between it and the material world' (p. 77)[50] – and
the point of distinction between man from animals (p. 40) – and once
recognised in this way, Bell maintained, 'the frame of the body,
constituted for the support of the vital functions, becomes the
instrument of expression'. Or, put slightly differently,

An extensive class of passions, by influencing the heart, by affecting that
sensibility which governs the muscles of respiration, calls them into cooperation, so that they become an undeviating and sure sign of certain
states or conditions of the mind. They are the organs of expression. (p. 88)

Represented by the muscles and the nerves of the body, these 'organs
of expression' provide a structure which links together the complex
network of relations that underlie the surface of the body for, Bell
claimed, 'it is not upon a single feature that the emotion operates;
but the whole face is marked with expression, all the movements of
which are consentaneous [sic] . . . the peculiar expression of individual emotion being distinguished by the action and determination
of certain features' (p. 140).

In assuming the role of instructor to artists, Bell placed himself
within the long tradition of using anatomy for artistic ends.[51] The
study of the antique, the living model, and anatomy had formed the
three main components of an artist's training, however informally,
since the Renaissance, yet the early academies in England at the end
of the seventeenth and the beginning of the eighteenth centuries,
and later the Royal Academy, had tended to neglect the study of
anatomy in favour of the study of the antique and the life class. It
was only with the constitution of the Royal Academy, in fact, that
the study of the living model and the antique were brought together
as subjects of comparative analysis which would prove crucial to the
development of an organised visual language of art. It is clear that
bodily examination, internal and external, was important to the
practices of art and medicine in the late eighteenth and early
nineteenth centuries.[52] But it was not until the early nineteenth
century that anatomy came to be seen as a practice of dissection,
observation, and drawing – rather than simply the reproduction of
plates from handbooks like those of Albinus and Vesalius, as Le

Brun's reproduced heads illustrate – and a legitimate alternative to the study of the antique. Observation of the living model was clearly invaluable to Bell's understanding of expression; nonetheless, he was all too aware that posed as it was, without reference to either the action of the muscles or the expressions of the face, the model presented an asymmetrical, constrained, and usually distorted version of the human body. The problem was how to represent the physical activities of everyday experience rather than the postures found in the Academy. Or, as Bell put it: 'When a man clenches his fist in passion, the other arm does not lie in elegant relaxation; when the face is stern and vindictive, there is energy in the whole frame; when a man rises from his seat in impassioned gesture, a certain tension and straining pervades every limb and feature' (p. 200). What the artist must learn is the structure of the bones and the groupings of the muscles so that he can 'observe attentively the play of the muscles and tendons when the body is thrown into action and attitudes of violent exertion [and] mark especially their changes during the striking out of the limbs' (p. 202). Bell's description of muscular motion in this, the third edition of *Anatomy of Expression*, seems to draw directly on Hartley's associationism, translating his explanation of involuntary movements into the realm of the aesthetic. Having learnt to draw the human figure in this way, Bell instructed, the artist should then make the model perform some actions – such as pitching, throwing, striking – so that he can advance his understanding of the 'character of muscular expression'.

The system of art education offered by the Royal Academy apart, Bell believed an ignorance of the principles of anatomy was a very real weakness for an artist and a sleight on the practice of painting. 'Suppose that a young artist, not previously grounded in anatomy, is about to sketch a figure or a limb', he suggested, directly attacking the Academy, 'his execution will be feeble, and he will commit many errors if he endeavour merely to copy what is placed before him—to transcribe, as it were, a language which he does not understand'. He explained at length:

He sees an undulating surface, with the bones and processes of the joints faintly marked; he neglects the peculiar swelling of the muscles, to which he should give force, as implying motion; he makes roundings merely; he is incapable of representing the elegant curved outline of beauty, with decision and accuracy, and of preserving at the same time the characters of living action. Drawing what he does not understand, he falls into tameness or deviates into caricature. (pp. 203–4)

An understanding of the language of anatomy, Bell insisted, provides a guarantee against this sort of error and misreading by teaching the art student to see the function and the structure underlying the surface and responsible for 'the peculiar swelling of the muscles . . . the elegant curved outline of beauty [and] the characters of living action'.

Benjamin Robert Haydon's enthusiasm for Bell's work was so strong that he attributed the completion of his studies in anatomy to his attendance at Bell's lectures in anatomy, surgery, dissecting and operating (delivered in his London house). An extremely dedicated student of Bell's, Haydon referred to his instructor in his *Autobiography and Journals* for that year, writing that he 'had great delight in his subject and was as eager as ourselves. Poor and anxious for reputation, he was industrious and did his best. He had studied and fully understood the application of anatomy to the purposes we wanted.'[53] To Bell, the use of the language of anatomy as an artistic technique provided a structural scheme to represent the expressions as well as a vocabulary to describe the movements of the features. His appeal to artists to develop a functional understanding of the human figure depended, therefore, on the construction of a grammar of expression which emphasised sensation and motion as evidence of the internal structure of the bones and the internal state of the emotions. Schooled by Bell to study the physical expression of the body from close quarters, Haydon accepted Bell's view of the body as a complex network of nerves and muscles. Haydon's appetite in these early years for 'Drawing – Dissection – and High Art' was so insatiable that he would spend hours every day pouring over plaster casts, anatomy books, and whatever parts of dead bodies he could get his hands on. Such was his enthusiasm that he sometimes lost sight of the object of study, with comic consequences. A year or so after he first came to London, for instance, Haydon arrived to take breakfast with a new acquaintance, David Wilkie, only to discover, to his acute embarrassment, that Wilkie was completely naked and utterly engrossed in drawing his exposed figure through a mirror. 'I went to his room rather earlier than the hour named', he said,

and to my utter astonishment found Wilkie sitting stark-naked on the side of his bed, drawing himself by the help of the looking-glass! 'My God,

Wilkie', said I, 'where are we to breakfast?' Without any apology or attention to my important question, he replied: 'It's jest capital practice!'[54]

Indeed, Haydon comments that having 'rallied' Wilkie on his 'copital practice', he 'shall never forget his red hair, his long lanky figure reflected in the glass, and Wilkie with port-crayon and paper, making a beautiful study'.[55] A year later, Haydon gushed excitedly at the experience of surveying a dissected body and its organs. 'The sight of a real body laid open', he exclaimed, 'exposed the secrets of all the markings so wonderfully that my mind got a new spring. The distinction between muscle, tendon, and bone was so palpable now that there could be no mistake again for ever.'[56] Later, an entry in his *Autobiography and Journals* for 1810 describes the process of dissection in erotic terms as an experience of vigorous and almost ecstatic pleasure:

While I was furiously at work on Macbeth, Charles Bell sent up to me to say that he had a lioness for dissection. I darted at it at once and this relieved my mind. I dissected her and made myself completely master of this magnificent quadruped. It was whilst meditating on her beautiful construction, and its relation in bony structure to that of man, that those principles of form since established by me arose in my mind.[57]

The sudden excitable movement by Haydon towards the animal – 'I darted at it at once' – and the immediate sensation of relief which it evokes – 'this relieved my mind' – suggests that the act of dissection involved both surgical exploration and erotic caress (to the artist at least). The tactile exploration of bodily form, that aligns the hand and the eye, appears to satisfy a personal and a professional desire for pleasure and knowledge and locates the experience of seeing inside the body.

Such was Haydon's aesthetic delight in the dissected parts of the body as evidence of its mechanical organisation, that it is not surprising that the training in his 'School' of art – formed around 1815 and numbering among its students Charles and Thomas Landseer, William Bewick, and Charles Eastlake – started with anatomical drawings and then progressed to dissecting cadavers, before finally drawing from the Elgin Marbles at the British Museum. What is more, it was to Bell's series of anatomy lectures that Haydon's students were sent to learn the theoretical and practical procedures for dissection, and to perfect the highly skilled methods of its representation, assisted, naturally, by that essential textbook, *Anatomy of Expression*. 'The consequence', Haydon remarked, 'certainly was a reform in the painting of the School, for though anatomy was considered a part of the study of the student it was not taken up as I took it up – thoroughly; and made my pupils do so'.[58] William Bewick's recollections of his studies at Bell's anatomy classes support Haydon's account; he recalls how he 'dissected at Sir Charles Bell's theatre of anatomy for three seasons with the Landseers. We dissected every part of the muscles of the body, and made drawings in red, black, and white chalk, the size of nature. These drawings were thought by the professor [Bell] the finest ever made from dissection.'[59]

The challenge for Bell was to identify the grounds upon which his physiognomical teachings could be established as true or challenged as false. Were they the result of empirical (scientific or artistic) observation? Were they conclusions derived from an anatomical understanding of the nerves and muscles? Was the artist obliged to begin from a thorough grounding in the knowledge of anatomy? Must the artist thus be obedient to science? For Bell it is evident that anatomy lays the foundations for the study of expression; that is to say, if the face is the canvas upon which the expressions of emotion are displayed by the actions of its different muscles, then anatomy is 'the observation of all the characteristic varieties which distinguish the frame of the body or countenance' (p. 194); it is 'the grammar of that language in which they [the expressions] address us':

The expressions, attitudes, and movements of the human figure are the characters of this language, adapted to convey the effect of historical narration, as well as to shew the working of human passion, and to give the most striking and lively indications of intellectual power and energy. (p. 2)

Through the study of anatomy, then, the artist will learn 'to observe nature, to see the forms in their minute varieties . . . to catch expressions so evanescent that they must escape him, did he not know their sources' (p. 2). To Bell's mind, anatomy enabled not just a comparison of the forms of nature but also an accuracy of observation in the process.

VI

At the heart of Bell's treatise on expression were three related axioms: a theory of the mind as an essence or spiritual entity whose activities were determined by the will of God; a belief that expressions corresponded to specific emotions, based on a physiological understanding of the nervous system; and the identification of specifically 'human' muscles for expression (see plate 8). The argument was that there is a beauty in expression which comes from real (not ideal) forms, the recognition of which involves distinguishing between the capacity of an individual to express emotion and the actual expression of emotion and thought:

A countenance may be distinguished by being expressive of thought; that is, it may indicate the possession of the intellectual powers . . . On the other hand, there may be a movement of the features, and the quality of thought, – affection, love, joy, sorrow, gratitude, or sympathy with suffering, – is immediately declared. A countenance which, in ordinary conditions, has nothing remarkable may become beautiful in expression. It is expression which raises affection, which dwells pleasantly or painfully on the memory. (p. 18)[60]

He proposed that the subject of art should be the 'ordinary conditions' of life, 'what is human . . . what stands before us, to be seen, touched, and measured' (p. 19). Though profoundly influenced by classical theories of beauty, Bell saw beauty in the variety of forms presented in nature: ideal forms appeared to undermine, and often erase, the 'truth of expression and character' because they were unable to express the momentary 'display of muscular action in the human figure' or 'the effect produced upon the surface of the body and limbs by the action of the muscles'. Thus, artists must draw the figures they see in everyday life, according to Bell, rather than selecting or imagining figures appropriate to a particular artistic convention; the reason was that 'beauty is consistent with an infinite variety of forms . . . its cause and origin is to be found in some quality capable of varying and accommodating itself, which can attach to different forms, and still operate through every change' (p. 181).

The basis for this philosophy of ordinariness was a belief that there was a correspondence between external appearance and internal structure: 'outward forms result from the degree of development of the contained organs' (p. 21). Bell reiterated this familiar proposition via the notion that expressions of the face present an index of mental development and, consequently, that character can be explained through an anatomical understanding of expression:

How much more beautiful is the picture when the anatomy is displayed, the thinness of a care-worn face, the ridge of the frontal bone highly

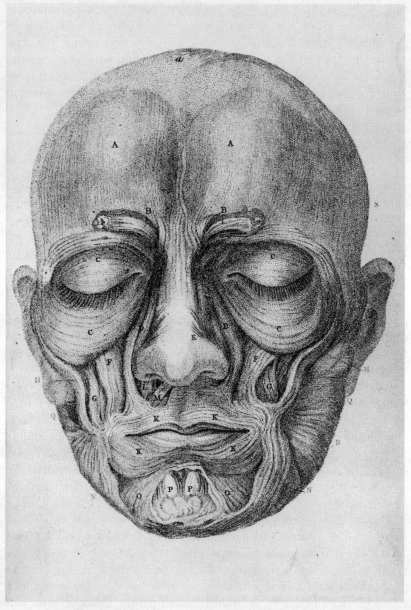

Plate 8 Charles Bell, The muscles of the face, 1877.

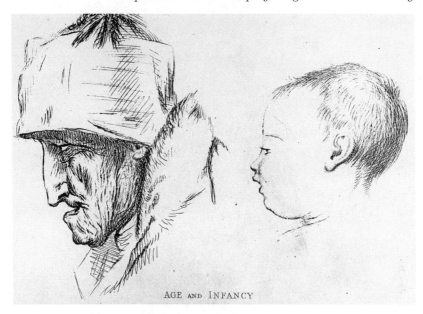

AGE AND INFANCY

Plate 9   Charles Bell, Age and infancy, 1877.

illuminated, the veins in their course over the temple, the delicate transparent colours of the skin, the shade of floating grey hairs. So much character will often be produced by the simplest touch presenting the true anatomy. (p. 8)

In fact, it helped the observer to develop 'a knowledge of the peculiarities of infancy, youth, or age; of sickness or robust health; or of the contrasts between manly and muscular strength and feminine delicacy; or of the appearances which pain or death present' (p. 194) (see plate 9). Thus, by focussing on the 'infinite variety of forms' in the physical world, Bell was able to provide a series of narratives for different facial expressions in terms of their distinctive muscular motions and emotional experiences. Laughter, weeping, and grief; pain, demoniacs, convulsions; despair and joy; admiration and jealousy; rage and remorse – these are just some examples of the expressions and emotions described by Bell (see plate 10). The important thing is that Bell derived uniformity from variety and, as a result, reduced the individuality of an expression to its typological form.

What was the method through which this reduction of particular

Plate 10   Charles Bell, Rage, 1877.

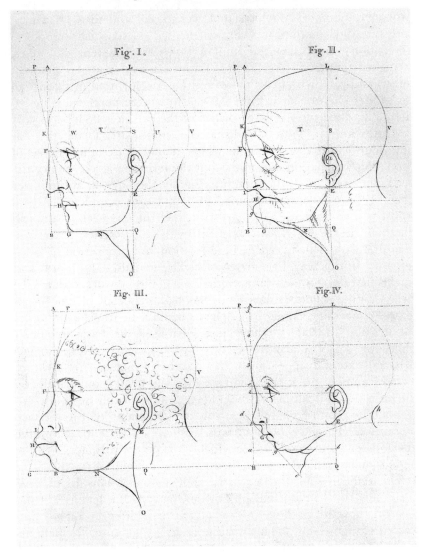

Plate 11   Pieter Camper, Proportional drawings of heads in profile, 1821.

(complex) to general (simple) worked? One of the most popular suggestions for a method which could be used to achieve this end was that the face could be measured against the head; as Bell pointed out, John Hunter, Camper, Blumenbach and Cuvier all entertained this idea. Of the four, Camper presented what is probably the most 'sophisticated' version of the idea. A Dutch surgeon and anatomist, Camper devised a statistical mode of perception which saw the human body as a system of circles, triangles, squares, and rectangles. To draw a head (plate 11), an artist must first draw two circles LVEW and KUZ, to form an oval, and an horizontal line ST. Then from the centre S, a perpendicular line SQ should be drawn to mark the orifice of the ear and the lobe at E. A line drawn from P to G, outside the circle KUZ, forms the all-important facial line or angle which demonstrates the degree of pronation/inclination and marks out the forehead K, the line of the eye F, the nose I, and the mouth H. Finally, the point Z, at the lower edge of the eye socket, is joined to V and E to complete the oval whilst a line from G to N marks the start of the neck. Camper's diagram of the head was formed by using a series of precise measurements which recorded the relation and proportion of facial features. In this way, the degree of civility and thus the moral and intellectual development of an individual could be worked out. Like Bell, Camper seems to have understood the concept of type in an almost Cuvierian manner as a defence against the transformation of the species. Indeed, we have just seen how the basis for Bell's philosophy of ordinariness lay in quite a specific correspondence between external appearance and internal structure. Yet Bell rejected Camper's attempt to map the face and head geometrically on the basis that it dealt primarily in ideal, antique forms (see plate 12). Instead, he proposed a new principle of expression: 'that in the face there is a character of nobleness observable, depending on the development of certain organs which indicate the prevalence of the higher qualities allied to thought, and therefore human' (p. 28).

The difficulty for the artist was not so much how to observe accurately the relation and proportion between the features of the face but how to visualise these features in relation to higher and lower emotional states. This was the crux of the matter for Bell: his belief in the mind as the spirit of an essentially higher being whose activities determined human action meant that the organs of the body were taken as connectors between mind and the external world; the body could influence the mind, and even under certain conditions help to develop states of mind, but it could never produce emotion, only mind could do that. 'By emotions', Bell declared, 'are meant certain changes or affections of the mind, as grief, joy, astonishment'. His goal, therefore, was to make manifest the corre-

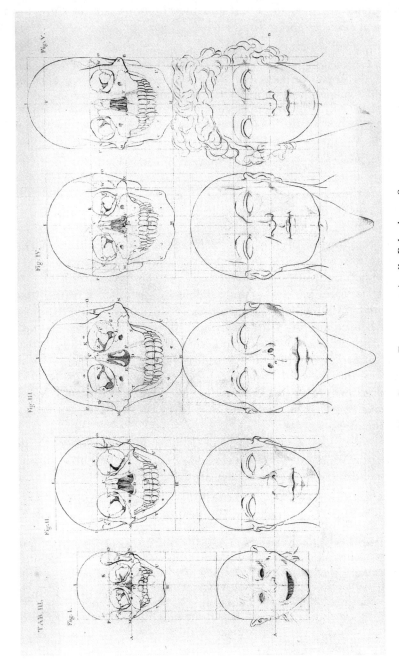

Plate 12   Pieter Camper, From ape to Apollo Belvedere, 1821.

spondence between the mind and the emotions, and to show from
where it emerges and to what effect:

What the eye, the ear, or the finger, is to the mind, as exciting those ideas
which have been appointed to correspond with the qualities of the material
world, the organs of the breast are to the development of our affections;
and that without them we might see, hear, and smell, but we should walk
the earth coldly indifferent to all emotions which may be said in an especial
manner to animate us, and give interest and grace to human thoughts and
actions. (p. 79)

But how were expression and emotion linked? What was the degree
of correlation between the movements of the face and the internal
state of mind? In Bell's understanding, the chain of events which
gives rise to the expression of emotion depends on the heart, the
organs of the breast, and its sensibilities, and inextricably connected
with the organs of the breast, he argued, were the organs of
respiration which develop and make visible the emotions. The
sequence might well have looked like this: an emotion disturbs the
heart, activates breathing, and produces expression. It was, in fact,
reasonably simple, for 'certain strong feelings of the mind produce a
disturbed condition of the heart; and through that corporeal influence, directly from the heart, indirectly from the mind, the extensive
apparatus constituting the organ of breathing is put in motion and
gives us the outward signs which we call expression' (p. 80). The
point is that any expression is the result of a compound activity
which emerges from the mind and involves the heart, respiration,
and the muscles.

Consider, for example, the expression of grief. 'So in grief', Bell
said, 'if we attend to the same class of phenomena [as a man of fear],
we shall be able to draw an exact picture':

Let us imagine to ourselves the overwhelming influence of grief on woman
[sic]. The object in her mind has absorbed all the powers of the frame, the
body is no more regarded, the spirits have left it, it reclines, and the limbs
gravitate; they are nerveless and relaxed, and she scarcely breathes [1]; but
why comes at intervals the long-drawn sigh? – why are the neck and throat
convulsed? – what causes the swelling and quivering of the lips, and the
deadly paleness of the face? – or why is the hand so pale and earthly cold
[2]? – or why, at intervals, as the agony returns, does the convulsion spread
over the frame like a paroxysm of suffocation [3]? (p. 82; my numbers)

Bell framed the type of grieving 'woman' within a physiological
model of expression which attempts to dissolve the distinction

between the external signs of grief and the internal anxiety of the emotion; or, in other words, to make grief visible from the inside and from the outside. For the grieving woman, the fraught activity of the mind appears to absorb all the muscular energy of her body until the point when this trance-like state is disrupted by the sporadic reflexions of the muscles. The neck and throat convulse; the lips swell and quiver; the face pales; the hand cools; and then a suffocating shudder sweeps through the body. In effect, it would seem there are three stages of grief: the mental absorption [1], the physical reaction [2], and a combination, even integration, of the mental and the physical [3]. The claim is that the expression of an emotion cannot be attributed solely to the activity of the mind; rather, the mind is the indirect source whilst the heart and lungs, 'and all the extended instrument of breathing' are the direct source of the expression of an emotion (p. 82). The heart and lungs have the same function in so far as their actions and motions are necessary to the circulation of the blood through the body, and so, Bell makes clear, the two organs are linked by the nerves and correspond (as action then excitement) accordingly.

The originality of Bell's thesis lies in this positioning of the nervous system as central to any understanding of the expression of emotions. Through his diagrammatic outlines of the muscles and the nerves of the face we can grasp the complexity and centrality of a specific class of nerves which are connected to respiration, so that its central nerve descends from the brain into the chest to link to the heart and lungs, and its other nerves link to the muscles of the chest, neck and face. Under the influence of the central nerve, it is the other nerves connecting to the chest, neck, and face which are the instruments of breathing and expression:

The heart and lungs, though insensible to common impression, yet being acutely alive to their proper stimulus, suffer from the slightest change of posture or exertion of the frame, and also from the changes or affections of the mind. The impression thus made on these internal organs is not visible by its effects upon them, but on the external and remote muscles associated with them. (p. 84)

Many claims could be advanced here, but Bell is in no doubt that the complexity of the nervous system, of which this class of nerves is exemplary, provides the best possible proof of the excellence of design: it confirms the mechanism which lies behind the 'common impression' and determines the means through which they are made visible. What is more, it supports Bell's belief that facial expressions correspond to specific emotions (based on precisely this physiological understanding of the nervous system).

We can see the scope and rationale for this correspondence between expression and emotion in Bell's description of blushing:

> The sudden flushing of the countenance in blushing belongs to expression, as one of the many sources of sympathy which bind us together. This suffusion serves no purpose of the economy, whilst we must acknowledge the interest which it excites as an indication of mind. It adds perfection to the features of beauty . . . Blushing is too sudden and too partial to be traced to the heart's action. That it is a provision for expression may be inferred from the colour extending only to the surface of the face, neck, breast, the parts most exposed. It is not acquired; it is from the beginning. It is unlike the effect of powerful, depressing emotions, which influence the whole body . . . the colour caused by blushing gives brilliancy and interest to the expression of the face . . . We think of blushing as accompanying shame; but it is indicative of excitement. There is no shame when lively feeling makes a timid youth break through the restraint which modesty and reserve have imposed. It is becoming in youth, it is seemly in more advanced years in women. Blushing assorts well with youthful and with effeminate features; whilst nothing is more hateful than a dog-face, that exhibits no token of sensibility in the variations of colour. (pp. 88–9)

The stages of blushing are less distinct than in the example of grief, mainly because it is considered more in terms of its effects than its cause; so, whilst it is connected to the mind, blushing is indicated by the suffusion of heat and colour over the most visible parts of the body (face, neck, chest), and arises from excitement and lively feeling. Blushing appears to be without purpose and, moreover, its expression seems to be unusual amongst the variety of expressions, according to Bell, as it does not involve muscular action, at least not directly. In this sense, blushing is easier to recognise but harder to explain than other expressions, such as grief, because it is involuntary and, more importantly, because the emotion it expresses is uncertain: it could be shame, excitement, shyness, modesty. Above all, blushing seems to be a socially conditioned response to a situation in which control appears to be lost. We will see in chapter five, when we consider Darwin's refutation of Bell's argument for expression, that blushing is one of the most important expressions to explain because it involves self-regulation in entirely physical terms without retaining mental command of a specific emotional outburst. For the moment, though, I want to look at one further example from

Bell's treatise which identifies the specifically 'human' muscles for expression and illustrates the extent to which his study mixes natural theological ideas with scientific explanations for the fit of expressiveness and its meanings.

Before Bell, it is safe to say that the prevailing assumption was that an 'intuitive faculty in the observer' provided 'special provision' in man for the expression of emotion and the subsequent interpretation of the relationship (p. 112); that much is obvious from Lavater's writings. However, Bell took a slightly different line, suggesting that whilst 'the provision for giving motion to the features in animals, and that for bestowing expression in man' were quite different, the important thing to recognise was the anatomical grounds for this difference which accounted for the existence of uniquely 'human' muscles designed solely for the purpose of expression (p. 113). The example of laughter (plate 13) bears this out. Bell claimed that animals are not aware of the sentiments which produce laughter: 'the capacity for receiving ludicrous ideas is as completely denied to animals as they are utterly incapable of the accompanying action of laughter' (p. 129). When we compare this to the description of laughter in humans, the claim becomes even stronger, as animals do not possess the muscular apparatus necessary to laugh. Upon laughing, the body experiences a number of motions:

The muscles concentring [sic] to the mouth prevail . . . they retract the lips, and display the teeth. The cheeks are more powerfully drawn up, the eyelids wrinkled, and the eye almost concealed. The lachrymal gland within the orbit is compressed by the pressure on the eyeball, and the eye suffused with tears [2] . . . Observe the condition of a man convulsed with laughter, and consider what are the organs or system of parts affected [1]. He draws a full breath, and throws it out in interrupted, short, and audible cachinnations [1]; the muscles of his throat, neck, and chest, are agitated; the diaphragm is especially convulsed [2]. He holds his sides, and, from the violent agitation, he is incapable of a voluntary act [3]. (p. 135; my numbers)

This was crucial to Bell's argument for expression, to prove not just the discontinuity of the species but also the function of what Bell termed the respiratory system of the nerves, increasing a fit between a specific emotion and its expression. Thus, if we examine the description of laughter we find there are three stages: the respiratory action [1], the muscular agitation [2], and a resulting state of involuntariness [3]. Moreover, these three stages from breathing and

Plate 13   Charles Bell, Laughter, 1877.

muscular contraction to an odd kind of paralysis dissect the experience of laughter according to physiological principles.

As I said at the beginning of this chapter, the value and significance of Bell's work is twofold: it affirms the importance of the divine aesthetics of the physiognomical tradition for the study of expression, and signals the import of new physiological doctrines in providing *scientific* rationale for the expression of emotions. The above examples of grief, blushing, and laughter confirm this assertion and reveal the intellectual pressures and tensions attendant upon those working within a natural theological framework. Of course, we had already glimpsed something of the demands of a teleological argument from Bell's essay on the brain and his treatise on the hand, but the real test for Bell was to outline the grounds upon which his physiognomical teachings could be established as true (or challenged as false). A knowledge of anatomy, in theory and in practice, served this end and, interestingly, makes evident the continuities and discontinuities in Bell's thought. The theory of expression which Bell presented actually became something of a classic in its time, in particular amongst the artistic and literary communities. Haydon was not unique in his enthusiasm for *Anatomy of Expression;* it was a well-known and widely used textbook for artists, anatomists, and writers alike throughout the nineteenth century, to the extent that Bell could count John Flaxman, Henry Fuseli, David Wilkie, and somewhat later Ford Madox Brown, John Everett Millais, William Holman Hunt, and Charles Darwin amongst those who read and admired his work, as well as Queen Charlotte and Princess Elizabeth, wife and daughter of George III respectively.[61] Furthermore, in the next chapter we will see one of the uses to which *Anatomy of Expression* was put by the Pre-Raphaelite Brotherhood, for whom the expression of emotion and character involved a truth which must come from ordinary people and their everyday experiences.

CHAPTER 3

What is character? The nature of ordinariness in the paintings of the Pre-Raphaelite Brotherhood

I

It is well known that the young artists who formed the Pre-Raphaelite Brotherhood in 1848 drew a great deal of their knowledge of the relation between anatomy and expression from Charles Bell's *Anatomy and Philosophy of Expression, as connected with the Fine Arts* (1844).[1] Given the increasing commercialisation of the practices of art and medicine, and the development of an informal exchange of techniques between the Royal Academy and the London medical schools, it is not surprising that a group of young artists should look to an anatomist for instruction. Bell's *Anatomy of Expression* became more or less a standard textbook with which young artists and their medical contemporaries were educated in the principles of anatomy and the practice of dissection; and in the case of the PRB, it provided them with a fulsome explanation of the advantages to be gained from surveying the physical forms in the organic world: '[anatomy] does not teach [the artist] to use his pencil, but it teaches him to observe nature, to see forms in their minute varieties . . . to catch expressions so evanescent that they must escape him, did he not know their sources'.[2] It is easy to see how appealing this kind of statement was: anatomy is an aid to knowledge, at once natural, visual, and physical, as it directs observation, assists comprehensions, and captures movement. What is more, it promised a means of achieving these things by attending to the real actions and events of the body. Hence, *Anatomy of Expression* was a much needed alternative to the limitations of the Royal Academy training, particularly with its emphasis on the study of the antique. William Michael Rossetti, one of the PRB's greatest advocates, explained their position:

Of the subjects recommended to our school as a body . . . the best, we think, are clearly those of our own day. But there is a distinction here. Mere

domestic art, as mostly understood and practised, is a very meagre affair
. . . boys playing games, girls listening to organ-grinders, cottagers smoking
quiet pipes, or preparing homely dinners. Or we have a touch of the most
poverty-stricken religious feeling – a grace before meal, or a girl at a
tombstone . . . Such Art as this is strictly analogous to the juvenile tale or
the religious tract; and it would be just as sensible to exhort our men of
letters to disport themselves in those mildest fields of literature as to inspirit
our painters to corresponding relaxations in art. Modern art, to be worthy
of the name, must deal with very different matter; with passion, multiform
character, real business and action, incident, historic fact . . . our great
Hogarth led the van of all modern-life art worthy of the name.[3]

Here, 'passion, multiform character, real business and action, incident, historic fact' stand in opposition to both the classical ideals
disseminated by the Royal Academy and the frivolity of contemporary art, and provide the backdrop against which the PRB developed
their aesthetic. They proposed an art of human nature in which the
emotions and experiences of everyday life were the focus of attention, rejecting idealised forms of religion and sentiment. By adopting
the techniques of the Italian Quattrocento painters and applying the
principles of anatomical expression, the PRB presented the organic
world with an enthusiasm and faithfulness which sometimes tended
towards severity and sharpness rather than what William Michael
Rossetti termed 'relaxations in art'. There were seven young artists
in all – John Everett Millais, William Holman Hunt, Dante Gabriel
Rossetti, James Collinson, Thomas Woolner, Frederick George
Stephens, and William Michael Rossetti – who founded the Brotherhood in 1848 and believed themselves capable of expanding the
range and depth of modern art by representing the variety and
changeability in nature.

Something of the ideological underpinnings of this conception of
nature – subsequently named a 'fuller nature' by Hunt[4] – can be
grasped if we look in detail at the origins and careful self-fashioning
of the PRB. There were four principles of PRB style, according to
William Michael Rossetti,

1. to have genuine ideas to express;
2. to study Nature attentively, so as to know how to express them;
3. to sympathise with what is direct and serious and heartfelt in previous
   art, to the exclusion of what is conventional and self-parading and
   learned by rote;
4. and most indispensable of all, to produce thoroughly good pictures and
   statues.[5]

Directed by these principles, together with a list of rules (probably drawn up by William Michael Rossetti) and a list of 'Immortals' – including such luminaries as Jesus Christ, Dante, Raphael, Shakespeare, and Keats[6] – the PRB presented themselves as aesthetic revolutionaries dedicated to recording the experiences which constituted ordinary life. They regarded the everyday world as a world of actual appearances which was mediated by the senses, the only means of grasping the real and the true. For Hunt, Millais, the Rossetti brothers, and the others, to express character involved capturing a truth which must come from the direct observation of life; life, that is, as proclaimed at dramatic moments of excitement and understood not as generalised appearance but inner spiritual and emotional experience. The four principles laid out by William Michael Rossetti pinpoint the detail of this approach because they identify the value of expression in painting as the means of communicating ideas of nature and also mediating prior artistic traditions. The main problem, though, was how to ensure that art served the ends of this life. For the PRB, as for Le Brun, the method of expressing emotion poses a considerable challenge to the artist because nothing could be more material than the expressions of the face. As Bell said: 'a countenance which, in ordinary conditions, has nothing remarkable, may become beautiful in expression'.[7] The contrast between the ordinary and the beautiful could not be more marked, and yet it is precisely this distinction that is elided by the artist in capturing expressions of emotion.

Through a highly imaginative grasp of history (derived in all probability from John Ruskin) the PRB constructed a dramatic art which is conceived in terms of the experienced character of feeling and found in the commonplace expressions of daily life. There is, they claim, an essential drama in life which drives human behaviour and colours our view of the world. The purpose of art was to capture this drama, almost directly imitating life and the characters expressed within it. Reality not ideality was the goal of the PRB aesthetic, which meant for them the ordinary conditions experienced by people on a day-to-day basis and in particular the emotions they felt in the process. What is the relationship between nature and character? How can character be expressed in terms both natural and dramatic? To what extent does character represent an internal (emotional) state? The PRB's approach to these questions will be the subject of this chapter.

A conviction that the proper subject of art ought to be found in the observation of the ordinary dominated the PRB aesthetic, for it was in ordinariness, they believed, that real beauty was found. Central to such an understanding was the belief that observation was the means of seeing through the individual to the common essences shared by individuals, gathering together the intuition of particulars into a representation of the substances which make us human. It was this process of seeing-as which translated observation into a synthesising, unifying activity that was essential to the development of what we now think of as a distinctively PRB aesthetic. The characteristics of this were simple – a faithfulness to nature, an attention to details, and a rigidly geometric sense of perspective – and the main task of the PRB artist was to describe and explain what was human, not divine, labouring to express 'what stands materially before us, to be seen, touched, and measured'.[8] The idea was to make nature real, lived, and human, to show it as that which was common to all things in the world and knowable through observation. Above all, this meant seeing the fundamental elements of human nature in everyday occurrences as evidence of the vital communion of man and nature.[9] Not for the PRB an art of ideal forms, but instead an art of brotherhood which made visible the extraordinary kinships contained within the ordinary. 'The name of our Body was meant to keep in our minds our determination ever to battle against the frivolous art of the day', Hunt pronounced, 'which had for its ambition "Monkeyean" ideas, "Books of Beauty", Chorister Boys, whose forms were those of melted wax with drapery of no tangible texture'.[10] Hunt is unequivocal in his criticism of contemporary painting (particularly of the sort produced by David Wilkie), with the result that the everyday is pitted against the domestic, surface against spirit, drama against theatricality. He presents the PRB (albeit self-consciously) as an avant-garde movement in paint precisely because of its rejection of these frivolities and flights of fancy.

The scope and rationale of the PRB's revolution are familiar to many.[11] They were determined to produce an art of the 'inner self' via the representation of individuals in action and thus convey the fundamental truths of human nature and life. For them, the expression of character posed an historical problem and an ontological challenge; that is to say, it confronted the artist with the academic assumptions and classical doctrines of previous artistic traditions whilst at the same time offering an opportunity to reformulate the nature of the individual as a subject caught in the world. We have already seen that the PRB turned to anatomy (and the rules of perspective) to meet the problem of tradition and the challenge of subjectivity – the latter was the more complex task as it required careful analysis of the conditions within which character emerges. Whilst anatomy quite literally opened up an individual to scrutiny, it also provided a means of observing the ordinary occurrences of life with a precision often otherwise lacking. The PRB noticed things, trivial actions and events, which eluded the attention of the casual observer; the danger was, of course, that attempts to render trivial things meaningful simply focussed attention on surface and action without substance. Just as for Lavater and Bell the complexity of nature, contained in the relation of mind to external world, was proof of the wisdom, benevolence, and power of the creator, so for the PRB the actual appearances of things in the world provided proof of the inner spiritual and emotional experience which lay under the surface. Hence, the decline of art was due to the 'tawdry glitter and theatrical pomposity' which, according to Hunt, characterised nineteenth-century English painting, whereas, on the contrary, the possibility of its rejuvenation and the recovery of its prestige depended on the expression of character, circumstance, and expression. 'Pictured waxworks playing the part of human beings provoked me', Hunt went on to say, 'and hackneyed conventionality often turned me from masters whose powers I valued otherwise. What I sought was the power of undying appeal to the hearts of living men. Much of the favourite art left the inner self untouched.'[12]

The claim was double-edged: not only was this an art which appealed to 'the hearts of living men' but it was also unconventional in so far as it attempted to touch the 'inner self' of the characters it sought to represent or, we might say, to capture the essence of an individual, the very core of their humanity. It was the theatricality of contemporary paintings that seems to have most vehemently offended the PRB sensibility. This is what Hunt had to say on the matter:

The fault we found in this younger school was that every scene was planned as for the stage, with second-rate actors to play the parts, striving to look not like sober live men, but pageant statues of waxwork. Knights were frowning and staring as none but hired supernumeraries could stare;

the pious had vitreous tears on their reverential checks; . . . homely couples were everywhere reading a Family Bible to a circle of most exemplary children; all alike from king to plebian were arrayed in clothes fresh from the bandbox. With this artificiality, the drawing was often of a pattern that left anatomy and the science of perspective but poorly demonstrated.[13]

Hunt's opposites in this passage are clear: 'pageant statues of waxwork' *versus* 'sober live men' and by implication, passive *versus* active, surface *versus* spirit. What emerges, in fact, is a subtle distinction between theatrical and dramatic expression, where the former is the worst kind of artifice and the latter is a natural form of expression visible in everyday life. An analogous representation of the 'inner self' can be found in Dante Gabriel Rossetti's later poem, 'The Portrait' (1868), which takes the face as exemplary of the character of an idealised female subject and transforms it into a *momento mori* for a lost love. The lines are as follows:

> O Lord of all compassionate control,
> O Love! Let this my lady's picture glow
> Under my hand to praise her name, and show
> Even of her inner self the perfect whole:
> . . .
> Lo! It is done. Above the enthroning throat
> The mouth's mould testifies of voice and kiss,
> The shadowed eyes remember and foresee.
> Her face is made her shrine. Let all men note
> That in all years (O Love, thy gift is this!)
> They that would look on her must come to me.[14]

Just as the poem conflates face with character, self, and soul, so Hunt equates theatricality of expression with superficiality and anatomy with movement. It is from this notion of the face that the activity of looking it dramatises became central to a PRB aesthetic, perhaps even talismanic in that it promised a way out of the strictures of the Academy style through an alternative school of thought.

An example of how this worked in practice is John Everett Millais' well-known painting *Lorenzo and Isabella* (1848–9) (see plate 14). Based on Keats' poem of the same name of 1818 and derived originally from Boccaccio, the painting offers an exemplary representation of an aesthetic designed to reveal the 'inner self' and functions as a means of propaganda, a special kind of manifesto, which identifies both the Brotherhood and Millais' affiliation to it by the placing of the initials 'P. R. B.' on the carved bench and the repetition of these

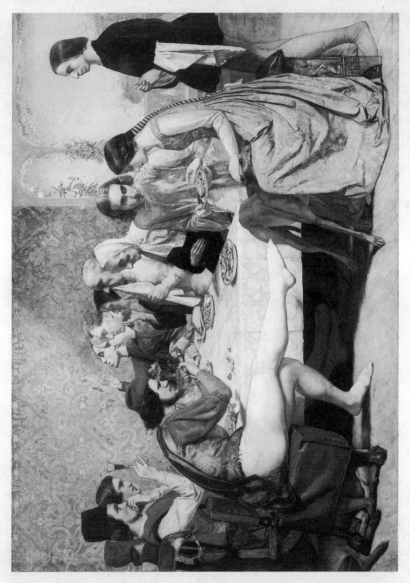

Plate 14 John Everett Millais, *Lorenzo and Isabella*, 1848–49

letters after his signature in the lower right-hand corner of the painting. Keats' poem tells the ill-fated story of two lovers – Isabella, daughter of a wealthy Florentine merchant, and Lorenzo, a clerk to the family business – Isabella's two brothers respond to her love for Lorenzo by murdering him, but having embarked on their danger-ous game of greed the brothers are surprised to discover that Isabella refuses to transfer her affections away from the memory of her absent lover. Indeed, prompted by a dream-vision, Isabella locates Lorenzo's body, cuts off his head, and plants it in a pot of basil. The brothers notice their sister's devotion to the basil plant, and on stealing the pot and realising its significance they depart from Florence in haste leaving Isabella to pine to death.

Millais' painting was accompanied in the Royal Academy cata-logue of 1849 with the following verses from an early part of the poem from which it was derived:

> Fair Isobel, poor simple Isobel!
> Lorenzo, a young palmer in Love's eye!
> They could not in the self-same mansion dwell
> without some stir of heart, some malady;
> They could not sit at meals but feel how well
> It soothed each to be the other by
> . . .
>
> These brethren having found by many signs
> What love Lorenzo for their sister had,
> And how she loved him too, each unconfines
> His bitter thoughts to other, well nigh mad
> That he, the servant of their trade designs,
> Should in their sister's love be blithe and glad,
> When twas their plan to coax her by degrees
> To some high noble and his olive-trees.[15]

Depicting members of Millais' family and some of his personal friends, the figures in the painting appropriate and reformulate the eighteenth-century tradition of portraiture as well as the conversa-tion piece popularised by Hogarth: they appear arranged in horizon-tal lines, dissecting the canvas with a range of terrifically reserved, intensely intimate, and portentously intrusive appearances.[16] This gallery of faces, arranged around the table in a style more like silhouettes than portraits, seems designed to draw the observer into the drama of the narrative as they scan the picture in a criss-crossing movement across the table and canvas. There is a strong sense of symmetry in the scene, but two movements, a kicking leg and a meeting of hands, disrupt a careful mapping of the visual field and unsettle its visible balance and order. Framed by two characters, a churlish brother with his white-stockinged leg kicking out in childish (and unmistakeably phallic) displeasure, and a meekly attentive servant, Lorenzo and Isabella share a blood-orange in a gesture of fatal tenderness. The plate on the table in front of the lovers, the two passion flowers on the balcony, the garden pots, and the hawk tearing at a white feather behind one of the brothers all symbolise the lovers' impending fate in highly dramatic terms. The contrast between the brothers and the lovers could not be more marked as each side of the familial divide is depicted in emotional form as anger and love respectively.

The focal point of the painting is Lorenzo's furtive glance at Isabella which communicates his devotion to her and at once seals his destiny. An early pencil sketch of the painting showed Isabella turning into the painting to look towards Lorenzo, but no such visual reciprocity of affection is evident in the final oil version wherein she stares almost vacantly down onto the table in front of her, literally transposing her affection from her lover to the grey-hound nuzzling at her lap. To the jealous gaze of the brothers, Lorenzo represents a dangerous intrusion into their family life, and more importantly their grand plan, but the effeminacy of his face betrays the vulnerability of his position: his close-set eyes, thin nose, small mouth, pointed chin, and fair hair all mark out his difference from the family members around him. What is more, he risks censure (and ultimately his life) to turn to face Isabella directly, becoming the only figure in the painting not to be viewed in profile and in so doing revealing his adoration for his beloved. It is as if the exhibition of his face detaches him from the action of his lover's life, condemns him to death, and thus removes him from the dramatic narrative. Lorenzo is constructed as a figure whose face is implicated in the process of representation, since his face, portrayed in the act of looking, displays the uncertainty of its occupation – poised between directness and indirectness or the gaze and the glance.

III

In William Michael Rossetti's opinion, the PRB artist had 'a determination to realise incident, and especially expression, from the painter's point of view – to make the thing as intense and actual as he could, quite careless whether the result would be voted odd, outré, horrid, frightful, and the rest of it'.[17] The significance of this statement could not be clearer: expressions, of a kind we make every moment of every day, contain the raw feeling necessary to make art come alive, injecting it with action and emotion and so rescuing it from the theatrical excesses of the style taught by the Academy. What is more, our response to such expressions is necessarily linked to the absorption and dramatic excitement of the individual experience as expressed on the face. Art should be able to copy life, ordinary life or life as it is lived, by translating these experiences into specific emotional expressions. And it is this movement from experience to expression which sets the aesthetic intentions of the PRB in direct opposition to the traditional assumptions about what constituted the proper subject of art.

'Fine art delights us from its being the semblance of what in nature delights us' claimed an article written for the PRB journal *The Germ* and signed by John Tupper (thought to have been Dante Gabriel Rossetti himself).[18] This article and its sequel make two related points:

[If] *Fine Arts* shall regard the general happiness of man, by addressing those attributes which are *peculiarly human*, by exciting the activity of his rational and benevolent powers . . . then the subject of Fine Art should be drawn from objects which address and excite the activity of man's rational and benevolent powers.[19]

Art, to become a more powerful engine of civilization, assuming a practically humanizing tendency . . . should be made more directly conversant with the things, incidents, and influences which surround and constitute the living world of those whom Art proposed to improve.[20]

Together these points indicate the purpose and direction of art, both in the feeling of pleasure it produces and the sense of social knowledge it offers. This sort of pronouncement echoes throughout the pages of *The Germ*, conceptualising and promoting the theoretical principles already familiar to many of the PRB artists. Frederick George Stephens outlined the relevance of an empirical approach to the arts, for example, maintaining that what can be seen should be known and, moreover, it ought to be displayed on canvas as the subject of art. He reasoned:

If this adherence to fact, to experiment and not to theory, – to begin at the beginning and not fly to the end, – had added so much to the knowledge of

man in science; why may it not greatly assist the moral purposes of the Arts
. . . Truth in every particular ought to be the aim of the artist . . . Admit no
untruth; let the priest's garments be clean.[21]

The stress in this passage is firmly placed on the intimate relation of
theory to practice; fact and experiment are, Stephens says, the tools
of the PRB trade and the means of moulding art into something
which helps people see what is familiar in an unfamiliar light. Even
more marked, though, is the inverted Platonism of the passage
whereby truth is seen in 'every particular' as the material foundation
of painting. It was precisely this emphasis on particularity, incident,
circumstance, and especially expression which produced the strongest criticism of PRB style, and perhaps, given their opposition to
the standards laid down by the Academy, it is hardly surprising.
Much of the resistance encountered by the PRB lay in the firm belief
of many contemporary art critics that there was a proper subject for
art but not in 'the minute accidents of their subject, including, or
rather seeking out, every excess of sharpness and deformity'.[22]
According to the most vociferous of critics, the expressions and
gestures of everyday life were too minor, incidental, and circumstantial for painterly representation, in fact too undramatic especially
when translated into religious, historical, and contemporary
scenes.[23] Most critics dispelled the careful explanations and justifica-
tions given by the PRB themselves and concentrated instead on the
rudeness of the physical forms they represented. For the PRB,
nothing in life which could be observed was undramatic and nothing
which could be dramatised should be left out of painting. By
constructing their paintings around what was most familiar in life,
namely actual appearances, the PRB promised the subject of art
would marry a world of values with a world of things. But to most
art critics the results were not only aesthetically disgusting but also
morally reprehensible.[24]

Anatomy was the main cause of contention. As Julie F. Codell has
argued, 'understanding the ties between anatomical language and
the symbolic meaning of that language as a social and moral nexus
of values may help to explain the vehemence of the critics' attacks
against the PRB depictions of faces and bodies'.[25] One example will
serve to illustrate the tenor of this type of response. Ralph N.
Wornum, an influential critic, took the opportunity to comment on
PRB style as an excuse to tirade in general and impassioned terms
against the deleterious effects of applying the principles of anatomy

to the practice of painting. 'When painting is the mere handmaid to morbid anatomy', he wrote, 'its path is clear and its duties fixed':

> it is then no longer Art, but an administrator to science, and it is without the pale of artistic criticism; but so long as painting is employed as an Art, its duty is to instruct and delight, certainly not to disgust . . . No exalted sentiment can possibly be aided by either ugliness or disease; . . . The physical ideal alone can harmonise with the spiritual ideal: in Art, whatever it may be in Nature in its present condition, the most beautiful should have the most beautiful body; lofty sentiment and physical baseness are essentially antagonistic.[26]

The gap between Wornum and Bell could not be greater: there is a total absence of fit between expression and beauty. Instead of art reflecting the perfections and imperfections of nature, we are offered a straightforwardly Platonic world of ideal forms which art exemplifies. Wornum's riposte against 'ugliness and disease . . . and physical baseness' is typical of a large number of negative reviews of PRB paintings, of which Millais' *Christ in the House of his Parents* (1849) and Hunt's *A Converted British Family Sheltering a Christian Missionary from the Persecution of the Druids* (1850) were signalled out for special ridicule.[27] A couple of aspects of this passage are notable: the first is Wornum's proposal that art and anatomy are incompatible because whereas art draws 'exalted sentiment' from the 'most beautiful body', anatomy concerns itself only with ugliness, physical baseness, even disease. The second is his refusal to see the physiognomic referent for PRB interest in expression and character which makes what happens on the face of things correspond to the internal state of individuals. The latter point is worth examining further because it suggests a hermeneutic problem that exists, then and now, in comprehending the real intentions of the PRB aesthetic.

What I am suggesting is that the very idea of expression in PRB painting assumes a certain kind of readability on the basis of its naturalness, the meanings of which are neither obvious nor transparent. That is to say, the PRB are very clear that the basis for their art lies in the ordinary conditions of life, and through dutiful attention to the various forms of nature that very ordinariness becomes extraordinary and exemplary. But exemplary of what? Mosche Barasch has recently argued that as a medium of communication, human expression and gesture 'must be divided into two essential parts: movements we believe to be part of "Nature", and movements based on a (more or less) deliberate use of available cultural conventions'.[28] He continues:

Gestures of the first kind are performed spontaneously, involuntarily, and perhaps even without our being aware of the fact that we are performing them. Blushing or turning pale, jerking back before a danger suddenly revealed to us – these are the examples we know from everyday life . . . The other type of gesture may be termed conventional. As 'symptomatic' gestures are derived from nature, the conventional gesture is considered as a product of what we call culture . . . Most important in our context is that conventional gestures are in the first place conceived as a means of communication. The readability of the natural gesture is a side- or after-effect; it has little to do with the aim of the gesture. The conventional gesture is – at least in its origins – performed in order to convey a message.[29]

Barasch's proposition is as follows: nature is to culture as symptomatic gesture is to deliberate gesture. He claims that those actions considered to be natural are not intended primarily to communicate but merely respond as involuntary movements. The distinction is, then, between communicative and reflex actions, and whereas in the previous chapters we saw Hartley's and Bell's use of the reflex concept, the PRB seem to devote themselves to the communicative value of gestures and expressions. Hence, what interests me in Barasch's argument is the way in which these two types of gesture, spontaneous and conventional, represent the internal and external mechanisms of response and so require different kinds of knowledge; the former physiological and the latter social. The PRB attempt to convey both these responses whilst appearing to emphasise the social over the physiological (or more accurately anatomical). The weakness of their style is due, therefore, to the difficulty of recognising the extent of their dependence on the actions and behaviour of social life. By looking in detail at Hunt's painting *The Eve of St. Agnes* (plate 15), it will become easier to see how this works itself out.

'The story in Keats's *Eve of St. Agnes*', Hunt claimed, 'illustrates the sacredness of honest responsible love and the weakness of proud intemperance', and he added, 'I may practise my new principles to some degree on that subject'.[30] A lengthy conversation between Hunt and Millais on the function of art, retold by Hunt, affirms their intention to represent 'living creatures' not 'waxen effigies'. As Hunt explained to Millais, '[you have] made beings of varied form as you see them in Nature. You've made living persons, not tinted effigies';

Plate 15   William Holman Hunt, *The Flight of Madelaine and Porphyro during the Drunkenness Attending the Revelry (The Eve of St. Agnes)*, 1848.

but he retorted playfully, 'that'll never do! It is too revolutionary.'[31] Successful in gaining acceptance for the 1848 Royal Academy exhibition, Hunt's painting was originally called *The Flight of Madelaine and Porphyro during the Drunkenness Attending the Revelry (The Eve of St. Agnes)* and was completed in just two months. 'Coming home at nine', he wrote, 'I worked on my canvas by the light of a lamp',

The architecture I had to paint with but little help of solid models, but the bough of mistletoe was hung up so that I might get the approximate night effect upon it; the bloodhounds I painted from a couple possessed by my friend, Mr J. B. Price; my fellow-student, James Key, sat to me for the figure of the sleeping page and for the hands of Porphyro, so I was enabled to advance the picture with but little outlay.[32]

Given Hunt's claims, it is worth noting that the head of the baron playing host in the background and the left hand of the porter were actually drawn by Millais in exchange for Hunt drawing part of the drapery around Iphigenia for his contemporary Academy piece, *Cymon and Iphigenia* (1848).[33] Hunt's painting was glossed in the exhibition catalogue with the penultimate stanza (XLI) of Keats' poem 'The Eve of St. Agnes':

> They glide, like phantoms, into the wide hall;
> Like the phantoms, to the iron porch, they glide;
> Where lay the Porter, in uneasy sprawl,
> With a huge empty flaggon by his side:
> The wakeful bloodhound rose, and shook his hide,
> But his sagacious eye an inmate owns:
> By one, and one, the bolts full easy slide: –
> The chains lie silent on the footworn stones: –
> The key turns, and the door upon its hinges groans.[34]

A tale of luxury and festivity, Keats' poems narrates the story of two lovers, Madelaine and Porphyro, estranged by familial decree but re-united on St. Agnes' Eve to break their forced separation and defy their ties of kinship in order to escape together. Yet, as Judith Bronkhurst has pointed out, the original and longer version of the painting's title, *The Flight of Madelaine and Porphyro during the Drunkenness Attending the Revelry* is an intentional misreading of Keats' poem, where the lovers escape only after 'the whole blood-thirsty race' have drunk themselves unconscious.[35] The slight retrospective shift in time from the poem to the painting allows Hunt to display a scene which aligns morality with expression in a manner which he would pursue, almost to its extreme, throughout his career.

Although Hunt concentrates on the scene of the lovers' departure at the end of the poem, the tension of this incredibly sensual, evocative poem, lavishly seductive and beguiling, is translated into the painting through the gestures and expressions of the main characters. Madelaine and Porphyro stand framed within the open door, symbol of their freedom and bond of love, and yet they occupy a liminal position on the margin between visibility and invisibility. The visual field of the painting is constructed through a complex series of physical movements: Madelaine's bare arm reaching across Porphyro's clothed body, and his hand stretching outwards to clasp the open door communicate the precarious vulnerability of their position caught in a moment of neither discovery nor escape. But while their glances are directed diagonally into the hall towards the stirring bloodhounds, their arms point away from and actually out of the hall. With the riotous festivities continuing unabated in the adjoining room, a sleeping page slumped in a chair and a porter slouched on the floor are placed in the foreground, surrounded by the debris of their evening, next to the stirring bloodhounds. By framing the painting in this way, the contrast between the collapse of the porter and page in their drunken stupor and the tentative, self-conscious standing posture of the lovers, frozen in anxiety and anticipation, emphasises the moral structure of the composition via physical gesture.

It is an important distinction of gesture that clearly indicates the difficulties identified by Barasch in representing physical movement naturally and spontaneously. The lovers' anxious glance across the room encourages a panoramic movement of the eye and compels the observer to become involved in the drama of the moment and interpret it in terms of the extended narrative of the lovers' intentions and resolutions. We have, then, a single frozen moment which stands in for the larger tableau-style narrative of Keats' poem, with its expansive actions and minutely detailed character, and reconfigures it as the converging space of hostility, rebellion, and most importantly, desire. Madelaine's face captures the observer's attention due to the subtle way in which she inclines towards the bloodhounds. The gentility and fragility of her features, especially her almond-shaped eyes, elegant nose, and beguiling lips, contrasts with the rapacious muscularity of the bloodhounds as well as Porphyro's downcast face and holds the action, guaranteeing the safety of their escape. With her wide-eyed glance almost but not quite facing out of the canvas, Madelaine represents an arresting figure whose apparently startled arm gestures belie the control she exerts on the narrative space; her face and the activity of looking it dramatises sustain the focus of the painting.

IV

In an important recent study on *Victorian Photography, Painting and Poetry* (1995), Lindsay Smith has discussed the influence of emerging photographic representation and its discourse on Pre-Raphaelite painting. Looking in particular at William Holman Hunt's *The Scapegoat* (1854–6) – now at the Lady Lever Art Gallery, Liverpool – Smith argues that the painting is recalcitrant to interpretation precisely because of its literality. 'Critics have variously considered the painting an aesthetic and iconographical failure', she says, 'primarily because of their persistent attempts to interpret the image in terms of a traditional realist mode':

For many viewers, the painting appears to provide an example of clear-cut representation; what we see is what we get – a goat displaying an exemplary treatment of hoof and hair. Contemporary reviewers, in particular, were at a loss as to how to begin to read it, not because they did not recognise the reference to Leviticus . . . but because the extent of its literality was such that they found it difficult to see beyond the fact that the painting represented a dying goat.[36]

Hunt's painting represents an episode from Leviticus wherein the Jews send the goat to death in the wilderness on the Day of Atonement as penance for their sins. But 'what does it mean', Smith asks, 'for the painting to be considered one of "a mere goat"? What does it imply to privilege its literality?'[37] These questions require an understanding of the complex set of issues which are central, she claims, to Pre-Raphaelitism and the aesthetic theory of John Ruskin and also William Morris. The emergence of photographic discourse introduces new challenges for the painter as well as new problems of interpretation for the critic (and observer) and helps to make sense, Smith suggests, of 'Hunt's decision to literalise the metaphoric':

as a consequence (in Ruskinian terms) the symbolic and imitative connotations in the painting radically intersect, producing wider ramifications for nineteenth-century visual theory'. It has to do with questions of the place of the transcendental in relation to the empirical, perhaps best understood in terms of the 'visible' – that which can be verified within the picture – and the 'invisible' – that which may only be inferred, but which is, none the less, reciprocally present.[38]

This is a notable thesis in that it makes manifest the difficulty of reading PRB painting and offers a plausible explanation for the attacks of critics like Wornum on the apparent crudeness of their work. It is not so much the readiness of critics (and observers) to read PRB painting in terms of its naturalness that is the cause of the difficulty, but that the meaning of paintings like *The Scapegoat* or *Isabella* or *The Eve of St. Agnes* are not transparent in these terms. The PRB are very clear, as we have seen, that the basis for their art lies in the ordinary conditions of life and the emotional responses it involves.

According to Ruskin, the emotional response to nature (and art) must be encouraged because its only function is a moral one; by rewriting the idea of the beautiful from a sensual, aesthetic perception of phenomena into a moral and theoretical conception, he allowed for a correspondence between nobility and ordinariness which we have seen became central to the PRB aesthetic. As he wrote in the third volume of *Modern Painters* (1853):

Physical beauty is a noble thing when it is seen in perfectness; but the manner in which the moderns pursue their ideal prevents their ever really seeing what they are always seeking . . . When such artists look at a face, they do not give it the attention necessary to discern what beauty is already in its peculiar features; but only to see how best it may be altered into something for which they have themselves laid down the laws. Nature never unveils her beauty to such a gaze. She keeps whatever she has done best, close sealed, until it is regarded with reverence. To the painter who honours her, she will open a revelation in the face of a street mendicant; but in the work of a painter who alters her, she will make Portia become ignoble, and Perdita graceless.[39]

Here, Ruskin impresses the importance of seeing the ordinary conditions of everyday life as the foundation of art. That is to say, nature will reveal its nobility to the artist who is willing to recognise the beauty which lies in the realm of real and lived experience: hence, art should reflect life not simply as appearances but as actual experiences. The point is that with a morality of art comes the duty or responsibility of the artist to portray the world accurately and faithfully. Such sentiments found their natural extension in Ruskin's letters to *The Times* (1851) in defence of the PRB. An apologist for PRB painting, Ruskin intervened on their behalf to explain the purpose of their painting and also to suggest the ways in which it should be interpreted. Obsessed with ensuring the validity of any enthusiasm for nature, he informed the readers of *The Times* that the PRB 'will draw either what they see, or what they supposed might have been the actual facts of the scene they desire to represent, irrespective of any conventional rules of picture-making'.[40] This distinction, between 'what they see' and 'what they supposed might have been the actual facts of the scene', is slight but significant as it pinpoints precisely the problem of literality identified by Smith. A subsequent letter from Ruskin to *The Times* clarifies the issue. 'Pre-Raphaelitism has but one principle', he said,

that of absolute, uncompromising truth in all that it does, obtained by working everything, down to the most minute detail, from nature, and from nature only . . . Or, where imagination is necessarily trusted to, by always endeavouring to conceive a fact as it really was likely to have happened, rather than as it most prettily might have happened . . . all agreeing in the effort to make their memories so accurate as to seem like portraiture, and their fancy so probable as to seem like memory.[41]

What horrified so many critics in the mid-century was the apparent absence of fit between the very literal representations of emotional expressions and the higher levels of explanation which art ought to provide. To critics such as Wornum, for instance, nobility in painting meant the expression of a superior conception of mind and body through ideal forms which in some way resemble expressions of emotion such as jealousy, love, or fear but not in their raw form.

   This notion of the ideal was drawn directly from the grand tradition of history painting as advocated by Joshua Reynolds, first president of the Royal Academy in the 1770s. Reynolds insisted on the importance of a number of oppositions: generality over circum-stantiality, the antique over the life class, sentiment over expression. The artist is counselled to develop a mode of representation that visualises the subject as an abstract and idealised figure who displays the universal qualities of a particular class, rank, 'sentiment [or] situation':

A painter of history shews the man by shewing his actions . . . He has but one sentence to utter, but one moment to exhibit . . . The painter has no other means of giving an idea of the dignity of the mind, but by that external appearance which grandeur of thought does generally, though not always, impress on the countenance; and by that correspondence of figure to sentiment and situation, which all men wish, but cannot command . . .

He cannot make his hero talk like a great man; he must make him look like one.[42]

Relying on action and appearance to figure the status of an individual, Reynolds' conception of history painting is communicated in a civic language of art which refers for its verification to 'grandeur of thought' as well as 'sentiment and situation', both of which define the terms of representation. The heroic and the beautiful are, therefore, situated in the proper sphere of polite aesthetic taste.[43] Opening his series of presidential lectures to the Academy, Reynolds addressed the duty of this 'great . . . learned . . . polite . . . commercial nation' through the function of the Academy as an 'ornament [to the] elegance and refinement' of the British nation. He explained: 'it is difficult to give any other reason, why an empire like that of BRITAIN, should so long have wanted an ornament so suitable to its greatness, than that slow progression of things, which naturally makes elegance and refinement the last effect of opulence and power'.[44] The purpose of history painting was to convey 'those expressions alone . . . which their respective situations generally produce' and also 'that expression which men of his rank generally exhibit'.[45] Of course, Ruskin shared some of Reynolds' opinions about art as the early volumes of *Modern Painters* reveal, in many ways seeking to do for nineteenth-century painting what Reynolds had done in the eighteenth century.[46] But he also supported the PRB, despite (rather than because of) their endeavour to move away from exactly these prescriptions laid down by Reynolds for the Academy.

We have already seen something of the difficulties attendant upon this move in discussing the response of critics to PRB painting. However, not all PRB artists or associates were as persuaded by the literal drama of life as Hunt. Ford Madox Brown, for instance, long-time supporter of the PRB aesthetic, placed great importance on the capacity of events to assume a typological or symbolic significance over and above the instinctive actions of an individual. Indeed, in a *Germ* essay entitled 'On the Mechanism of a Historical Picture' (1850), Brown proposes a very specific reformulation of Reynolds' notion of history painting.[47] Counselling the artist on the problems of 'giving body to his idea', he explained the importance of providing a material frame for the composition:

Having, by such means, secured the materials of which his work must be composed, the artist must endeavour, as far as lies in his power, to embody the picture in his thoughts, before having recourse to paper. He must patiently consider his subject, revolving in his mind every means that may assist the clear development of the story: giving the most prominent places to the most important actors, and carefully rejecting incidents that cannot be expressed by pantomimic art without the aid of text. He must also, in this mental forerunner of his picture, arrange the 'grouping' of his figures, – that is, the disposing of them in such agreeable clusters or situations on his canvass [*sic*] as may be compatible with the dramatic truth of the whole, (technically called the lines of a composition). He must also consider the color [*sic*], and disposition of light and dark masses in his design, so as to call attention to the principal objects, (technically called the 'effect').[48]

The conception of art given by Brown in this passage involves the dramatisation of a particular moment through careful attention to the internal mechanism of the narrative on display. This 'panto-mimic art without the aid of text' appears to invoke the grand tradition of history painting defined by Reynolds whilst at the same time rewriting the terms of its representation. Not for Brown was painting intended to idealise individual actions and gestures, but instead it was meant to capture the real drama of its subject as intensely and, most importantly, as accurately as it was experienced and felt.

The significance of the experienced character of feeling is often overlooked in painting, as Michael Fried maintains in a study of eighteenth-century French painting which is instructive for its theorisation of experience. In *Absorption and Theatricality* (1980), he examines French paintings of the period between 1753 and 1781, focussing in particular on the work of Denis Diderot in order to outline a turn towards an overtly dramatic conception of art.[49] An emphasis on representations of the character of absorption as 'a lived condition or mode of being' is, he says, a new and important component of the painting of this time.[50] This conception of painting as absorptive tended to involve the representation of individual action and emotion, directing attention towards an individual engrossed in an activity as if oblivious to an audience:

In sum, for Diderot and his contemporaries, as for the Albertian tradition generally, the human body *in action* was the best picture of the human soul; and the representation of action and passion was therefore felt to provide, if not quite a sure means of reaching the soul of the beholder, at any rate a pictorial resource of potentially enormous efficacy which the painter could neglect only at his peril.[51]

To Fried, the paintings of Chardin epitomise this concept of absorption because there the absorption of the human body in action 'strikes us not only as an ordinary, everyday condition but as that condition which, more than any other, characterises ordinary, everyday experience: as the hallmark or *sine qua non* of the everyday as such'.[52]

It is clear that the PRB aimed to achieve exactly this characterisation. Hunt, we recall, distinguished between theatrical and dramatic expression, rejecting the former as artifice in favour of the naturalness of the latter. And in a pronouncement which echoes Hunt's idea of the drama of life, Fried states: 'the primary or *dramatic* conception calls for establishing the fiction of the beholder's nonexistence in and through the persuasive representation of figures wholly absorbed in their actions, passions, activities, feelings, states of mind'.[53] Given the strength of Hunt's opposition to the Academy (and to 'Sir Sloshua' as Reynolds was called by the PRB), the likely (if indirect) sources for this dramatic conception of art were most probably the acting handbooks and manuals available at the time. It is unclear whether the PRB read these (or other) acting manuals, but there is a contiguity of thought with them which is especially evident in the work of Hunt, Millais, and Brown. It is clear from the discussion thus far that the PRB had a pronounced interest in the drama of life and its fitness as a subject of art.

Two of the most popular practitioners of dramatic theory were Henry Siddons and George Grant, both working in the first half of the nineteenth century but adopting different strategies to enlarge upon the theory and practice of acting. Siddons' *Practical Illustrations of Rhetorical Gesture and Action* (1807) illustrates an account of pantomime, in particular, as a rhetorical art of expression directed by intricate patterning and transformations.[54] By observing nature and recognising its design in relation to art, the actor will be able to express the variety of the passions in a wholly realistic light, according to Siddons. He stressed the mutual concern of both drama and painting with the physicality of action accessible to the observer via a distinctive series of gestures and expressions which he termed pantomimic. A pantomime, he says, represents an action and interests its audience through, 'the modifications of the body, which depend upon the cooperation of the soul'.[55] A pantomimic language is, by its nature, predicated on visual not verbal signs and is performed through physical action as a 'play of gesture' which signals 'the expressions of the different situations of the soul'.[56] An earlier passage makes clear that this 'play' functions as a mode of mimicry that refers the visible movements of an individual to the invisible motions of the soul:

If the gestures are exterior and visible signs of our bodies, by which the interior modifications of the soul are manifested and made known, it follows that we may consider them under a double point of view: in the first place, as visible changes of themselves; – secondly, as the means indicative of the interior operations of the soul.[57]

There are a couple of points worth noting here: the first is the adoption of a Cartesian model of mind and body (as used by Le Brun); and the second is the equal application of this statement of the value of an understanding of gesture to the different art forms of drama and painting. Aligning artistic with dramatic terminology, Siddons describes different kinds of gestures: those that are 'visible changes of themselves' are 'gestures picturesque', while gestures that are the 'means indicative of the interior operations of the soul' are 'gestures expressive'.[58]

Whereas Siddons sees the point of a shared language between drama and painting, Grant's *Essay on the Science of Acting* (1828) provides a more abstract reading of the propriety of dramatic self-presentation.[59] Affirming the link between painterly and theatrical modes of expression, he claimed that 'there is an art of colouring peculiar to dramatic writing, which, though in many respects it may be different from that in painting, yet it is to be conducted by the same rules':

We require of both the same strength of tint, and the same distinctions in the distribution of the brightness and shadow, the same caution in observing the gradation of lights, and the same art in throwing objects to a distance, or in bringing them immediately under the eye.[60]

The fact that the PRB rejected the Academy models in favour of ordinary subjects of the kind referred to by Ruskin, for instance people you might pass in the streets of London, suggests a dramatic understanding of the life of an individual akin to what Siddons and Grant insinuate. To be sure, the face speaks for the drama of life as it has a history which conditions the representation of its emotional expressions – the face communicates both the past life of an individual and the current circumstances of that life. These circumstances might consist of either the actions of an individual engaged in a specific task or nothing more than a single glance or gesture. To render expression involved capturing a truth which was only present in everyday life and could only come from the direct observation of actual, verifiable experiences. Yet, as I have shown, the expressions and gestures of the everyday were all too often deemed rather minor, incidental, and circumstantial to be the proper subject of painting.

v

It was the very idea that ordinary expressions were too undramatic (and so uninspiring) to be a fit subject that had the greatest currency in the critical debates of the mid-century: metaphor ought to remain metaphoric rather than be literalised in the hands of the PRB. Hunt, however, insisted on the drama in everyday life which ought to be conveyed in painterly form. He seemed to be looking for a natural mode of representation which encapsulated those gestures and expressions termed 'symptomatic' by Barasch. Instead of an opposition of nature and culture, therefore, Hunt proposed an alignment of nature and culture against artifice. A strong image of the theatricality he so despised is evident in a number of contemporary articles which prescribe ways of reading metropolitan culture in rigid theatrical terms.

'We have no need to go abroad to study ethnology', Eliza Lynn asserted in an article on 'Passing Faces', published in Charles Dickens' weekly journal *Household Words* in 1855, for 'a walk through the streets of London will show us specimens of every human variety known'.[61] 'It is', she continued, 'perfectly incredible what a large number of ugly people one sees',

one wonders where they can possibly have come from, – from what invading tribe of savages or monkeys. We meet faces that are scarcely human, – positively brutified out of all trace of intelligence by vice, gin, and want of education; but besides this sad class, there are the simply ugly faces, with all the lines turned the wrong way, and all the colours in the wrong places.[62]

Considered as a racial and zoological spectacle, 'exhibiting specimens of every human variety', the streets of London present the discerning eye and 'educated perception' with a threatening panorama of 'scarcely human' faces which counter received standards of judgement. Lynn identifies the lines and colours of certain individuals as the distinguishing marks of identity in the social milieu

and, she explains, though the artist alone is trained to 'trace the
original lines through the successive shadings made by many generations of a different race', a shrewd observer can nevertheless
recognise 'those lines . . . seen by all who know how to look for
them, or who understand them when they are before them'.[63]
Motivated by a desire to fix identity in terms of prevailing characteristics, Lynn articulates a peculiarly Victorian obsession with preserving the authenticity and integrity of the self through an exceedingly
narrow framework of lines and colours. Caught amidst the visual
culture of London which Lynn portrays, the perceptual clues to the
history, spiritual condition, and by implication class of an individual
are found in the numerous 'passing faces'. She evokes the endless
activity, changeability, and fluidity that daily confronts the dwellers
in the metropolis like a theatrical display; the reader is directed as
follows:

> Past the Circus – up Regent Street . . . – through Oxford Street, and
> towards Marble Arch – crowds upon crowds still meet; and face after face,
> full of meaning, turned towards you as you pass; signs of all nations and
> races of men pass you, unknown of all and to themselves whence they
> came; beasts and birds dressed in human form; tragedies in broadcloth,
> farces in rags; passions sweeping through the air like tropical storms, and
> silent virtues stealing by like moonlight; LIFE, in all its boundaries, power
> of joy and suffering – this is the great picture-book to be read in London
> streets, these are the wild notes to be listened to; this is the strange mass of
> pathos, poetry, caricature, and beauty which lie heaped up together
> without order or distinctive heading, and which men endorse as Society
> and World.[64]

It is a compelling description: the representation of life as a 'picture-
book' suggests the availability of models upon which the figures can
be read. The streets of London are aligned with the theatre as a
result of their corresponding displays of extreme emotions and
spectacular characters: 'beasts and birds dressed in human form;
tragedies in broadcloth, farces in rags; passions sweeping through
the air like tropical storms', all of these embody contemporary
theatrical types and behave in easily recognisable ways in the
kaleidoscope of life. The faces in the crowd are codified as personi-
fications of legible types of expression and, as a result our attention is
drawn to a theatricality which may be inherent, or at least implicated, in the everyday experience of life.

Writing a review article for the *Quarterly Review* a few years earlier,

Elizabeth Eastlake anticipates Lynn's sentiments as she enthusiastically describes 'the tremendous responsibility given to the human countenance, in the social economy of the world, as the great medium of recognition between man and man'.[65] The face is not only the 'badge of distinction' and the 'proof of identity', according to Eastlake, but also 'the sole proof which is instantaneous – an evidence not collected by effort, study, or time but obtained and apprehended in a moment'.[66] And as such it was a means of self-preservation which helped to counter 'the most bewildering confusions and fatal mistakes' by analysing character within a marked social hierarchy: 'it is the spirit within witnessing . . . with the spirit of the gazer, which alone touches the electric springs of recognition'.[67] For, as Eastlake is quick to point out, the modesty and decorum which can be seen on the face depend on fitting a particular face to its general type. The noblest were, then, always the most beautiful and the most moral whilst the least noble were consigned to being ugly and immoral: thus, 'every sex and every age of life has a physiognomy proper to itself, and only to be rightly defined by its dissimilarity to that of another. Each has a beauty after its kind, which it belongs to the true artist to observe and to the true physiognomist to discriminate.'[68] Like Lavater, Eastlake is advocating an essentialist view of human nature whereby each individual has a physical form which is 'proper' or suitable to their social (and spiritual) position in life. It is left to the physiognomist to identify the differences in kind and the artist to capture it on canvas.

A physiognomical handbook on the connection between 'linear and mental portraiture morally considered, and pictorially illustrated' written by Thomas Woolnoth in 1852 stresses the fundamental importance of seeing difference.[69] Indeed, Woolnoth creates the image of a peripatetic observer of the face as a collector of expressions after the manner of a natural historian:

In walking the streets of the metropolis, we have the finest opportunities of enlarging our facial observations: for in such a collection, all the expressions seem brought together as though for immediate comparison; hence we find in the great multitude the mixed Expression is the prevailing one, and has that neutralising effect upon the mass, that they move on as undistinguished as if they had no Expression at all . . . What arrests the eye in passing is that more turbulent and depraved condition of face, which does not average above one in a hundred, of such as are not to be brought suddenly or severely under physiognomical survey.[70]

The streets of London proffer a collection of mixed expressions as specimens for physiognomical analysis and classification; it is, however, the face which is not mixed but exceeds those exhibited by most people which strikes and perplexes the eye of the enlightened observer. Because, as Woolnoth makes obvious, most expressions of the face are classified empirically according to type but there are some which resist classification and instead fall outside the physiognomic mode of interpretation, corrupting the standards of expression. At issue are the ethnic origin, social status, and spiritual character of an individual's face and, perhaps more importantly, the readability of its features according to the physiognomical survey described by Woolnoth.

VI

The paintings of the PRB confront us with a representation of life wherein the expression of emotion, its mode and manner of presentation, is the subject of study. A number of paintings, including Millais' *Lorenzo and Isabella* and Hunt's *Eve of St. Agnes*, attempt to represent the emotional expressions of ordinary life as individual rather than typical configurations. As the discussion of some contemporary journal articles has demonstrated, an interest in the painterly representation of ordinary people tended, usually, to involve a typological form of symbolism. Similarly, in the catalogue for the 1865 exhibition of his paintings, Brown said, 'my object . . . in all cases is to delineate types, not individuals'.[71] His statement is interesting as it seeks to see in the particulars of an individual the manifestation of a common type, so reducing complexity to the simplicity of a limited number of representational forms. Describing his painting *The Last of England* (1852) in the same catalogue, he clarified his sense of the problem of reading literally as opposed to metaphorically. 'This painting is', he said, 'in the strictest sense historical':

it treats of the emigration movement which attained its culminating point in 1852. I have, therefore, in order to present the parting scene in its full tragic development, singled out a couple from the middle classes, high enough, through education and refinement, to appreciate all they are now giving up, and yet depressed enough in means, to have to put up with the discomforts and humiliations incident to a vessel 'all one class'.[72]

The narrative explains the present situation of the voyagers in terms of their emotions and so presents a vision in which any sense of pictorial unity is derived from a shared experience of loss and parting which brings together the various characters as a single type by virtue of there only being 'one class'. Attending closely to the dramatic potential of the painting, Brown went on:

The husband broods bitterly over blighted hopes and severance from all he has been striving for. The young wife's grief is of a less cankerous sort, probably confined to the sorrow of parting with a few friends of early years. The circle of her love moves with her . . . Next to them in the background, an honest family of the green-grocer kind . . . Still further back a reprobate shakes his fist with curses at the land of his birth, as though that were answerable for his want of success; his old mother reproves him for his foul-mouthed profanity, while a boon companion, with flushed countenance, and got up in nautical togs for the voyage, signifies drunken approbation. The cabbages slung round the stern of the vessel indicate to the practised eye a lengthy voyage; but for this their introduction would be objectless.[73]

This description endows the faces of husband and wife with a value as a specific type of social class, and yet it is the intuition of particulars from the faces of the other voyagers – a rogue, his old mother, and drunken companion – which typifies the drama of the moment.

Mary Cowling's study takes Brown at his word, suggesting that the popularity of physiognomy derived from its capacity to employ typological forms of classification, and as a result it became an important resource for and was often appropriated by mid-nineteenth-century genre painters.[74] But the introduction makes clear that Cowling's work simplifies many of the complications involved in representing character through expression. There is no doubt that physiognomy was a form of hermeneutics which was used frequently in artistic and literary contexts as an important means of character-ising subjects, as Cowling claims. Nonetheless, from Lavater to the PRB, each and every attempt to read and judge character was a means of ascribing an essence to human nature, visualising some-thing hidden from external appearances which, once revealed, made them both purposeful and substantial. An anonymous article in the *Quarterly Review* on 'The Physiognomy of the Human Form' (1856) addresses the value of an ordinary life as the subject-matter for painting and in so doing indicates something of the difficulty of what was involved in expression.[75] 'To symbolise is not, indeed, the chief or primary object of the construction of these parts [of the body]', the author wrote, 'but neither is it so of any of the features of the face':

The general law of symbolical construction is that forms are made to be significant without interfering with the fitness of the parts for other purposes than those of symbolising. The body and mind, the sign and the thing signified, do not correspond as effect to cause, but as things derived from a common origin, and planned with one design. They are in no relation of sequence . . . but . . . there is a perfect congruity between them; the body is the image of the mind, and, in the visible, the invisible is revealed.[76]

The author suggests that there are, in fact, four different types of symbols which can represent the human form – the general (masculine and feminine), the intellectual, the organic (physical shape), and the transient – and it is the ability to recognise and interpret these symbolic types that enables the observer to understand character. Judging character from physical form by seeing the invisible in the visible is, the reviewer says, an essential means of comprehending the variety in the 'natural pantomime of life';[77] moreover, this skill can be practised simply by applying the four types to the people we come across in everyday life. The implication is that far from being an artificial means of capturing expression, this attention to individuals as natural symbols allows most observers to recognise the extraordinary significance contained in that most ordinary of forms, the human face.

This chapter has explored the ways in which the PRB mediate the individual with the type, replacing an idealised (and idealising) form of representation with a dramatic conception of painting which sought to use expression and gestures as a means of conveying the often ordinary circumstances in which sharp pangs of emotion, in pain or pleasure, are experienced. Taking their lead from Bell, who saw the potential for beauty in the expressions of an ordinary face, the PRB in general, and Hunt in particular, advanced an aesthetic which promoted the profound and essential nobility in ordinary, everyday actions, expressions, and gestures; in effect, those things which make us human. Something akin to what Fried termed absorption, the PRB proposed the visibility of an invisible, internal state through emotional expression, and, to the horror of many a critic, they laid great emphasis on the experienced character of feeling. The literalisation of the metaphoric which occurs in many PRB paintings poses a particular problem of interpretation as it invites a reading in terms of typology and symbolism on the one hand, and rejects it on the other. The weakness of the PRB aesthetic lies in their inability to fully work their way out of this conundrum between type and individual; that is why Brown's statements and the notion of a dramatic form of art are fraught with contradictions, at times redolent of what they appeared to have rejected.

CHAPTER 4

'Beauty of character and beauty of aspect': expression, feeling, and the contemplation of emotion

An emotion taken in its whole range is a highly complex thing. It may begin in some local stimulus, as in a sensation of some of the senses, but it creates along with the feeling or the conscious state, a wave of diffused action of muscles, secreting organs, &c., including gesture, expression of features, and utterance, – all which become incorporated as part and parcel of the phenomenon.

Alexander Bain[1]

I

The idea that superior physical beauty was the expression of higher mental development was quite commonplace in the mid-nineteenth century. In an essay on 'Personal Beauty' (1854), Herbert Spencer sought an illustration of the natural congruity of 'beauty of character and beauty of aspect'.[2] His claim that in the process of sexual selection a 'good physique' was at least as important as a woman's 'moral and intellectual beauties', was driven by a sense of women's peculiar place in nature.[3] As Spencer increasingly recognised, the physical laws of development did not necessarily correspond to the moral laws which were responsible for regulating the life of an individual; any cultivation of the mind at the expense of body would, he claimed, be detrimental to the position of woman in the natural order:

The truth is that, out of the many elements uniting in various proportions to produce in a man's breast that complex emotion which we call love, the strongest are those produced by the physical attractions; the weakest are those produced by intellectual attractions; and even those are dependent much less upon acquired knowledge than on natural faculty – quickness, wit, insight. If any think the assertion a derogatory one, and inveigh against the masculine character for being thus swayed; we reply that they know little what they say when they thus call into question the Divine ordinations . . . It needs but remember that one of Nature's ends, or rather her supreme end, is the welfare of posterity . . . it needs but to remember that, as far as posterity is concerned, a cultivated intelligence based upon a bad physique is of little worth, seeing that its descendants will die out in a generation or two.[4]

It is interesting that Spencer's attempt to prove the correspondence of plainness with imperfection and beauty with perfection has marked similarities with the criticisms of PRB art; hence the familiarity of some of the sentiments of this passage. To see nobility in plainness or ordinariness in beauty was the key to PRB art, but advocating this sort of connection was to run the risk of censure, not to mention ridicule. So, whereas 'plainness may coexist with nobility of nature', in general the process of induction led Spencer to claim that 'mental and facial perfection are fundamentally connected, and will, when the present causes of incongruity have worked themselves out, be ever found united'.[5]

The assertion of a direct relationship between the features of the face and the faculty of mind is, of course, a reassertion of the meanings of expression embodied in earlier traditions in that it amounts to a classic statement of the physiognomic understanding of character. But what perplexed Spencer was how to explain the relationship between mind and facial expression when the congruity of character and aspect was not self-evident. He compared the babies of humans to those of animals, for instance, as well as the appearance of ugly people to 'inferior races':

If the recession of the forehead, protuberance of the jaws, and largeness of the cheekbones, three leading elements of ugliness, are demonstrably indicative of mental inferiority – if such other facial defects as great width between the eyes, flatness of the nose, spreading of the alae, frontward opening of the nostrils, length of the mouth, and largeness of the lips, are habitually associated with these, and disappear along with them as intelligence increases, both in the race and in the individual, is it not a fair inference that all such faulty trials of feature signify deficiencies of mind?[6]

It is well-known that Spencer's interest in mind and character was motivated by social concerns such as the origins of morality, the construction of educational programmes, and theories of progress.[7] The theory was that the progressive model of intellectual growth alluded to by Spencer in these sentences could be mapped quite literally through the profile and dimensions of the countenance.

What is more, the defence of his belief in the 'organic relationship between that protuberance of jaws which we consider ugly, and a certain inferiority of nature'[8] took the form of a profound sense of the intellectual disparities between men and women. This sexual division of labour was seen as the necessary consequence of sexual function and assumed, as Nancy Paxton has observed, that 'woman's most important contribution to the evolution of the race was a healthy "physique", which her education could cultivate, while man's contribution was a fully developed brain which his schooling, likewise, should foster'.[9]

An analysis of feeling which aligns the mental (or psychical) with the physical and character with aspect renders beauty an index of intellectual development. But how far should the divine aesthetics of the physiognomical tradition be seen to underpin these concerns with healthy physiques and beauty? Sander Gilman has discussed the ways in which conceptions of the ugly ('diseased') and the beautiful ('healthy') sustain social orders through biological narratives. The ugly, Gilman claims, is 'anti-erotic rather than merely unaesthetic. It is denied the ability to reproduce'.[10] The point is that placed in the context of natural and sexual selection, notions of good looks and mental capacity were inextricably linked to the fitness of women to reproduce. 'I may cite the passions which unite the sexes', Spencer wrote:

this is habitually, but very erroneously, spoken of as though it were a simple feeling; whereas it is in fact the most compound, and therefore the most powerful, of all the feelings...Thus, round the physical feeling forming the nucleus of the whole, there are gathered the feelings produced by personal beauty, that constituting simple attachment, those of reverence, of love of approbation, of self-esteem, of property, of love of freedom, of sympathy. All these, each excited in the highest degree, and severally tending to reflect their excitement on each other, form the composite psychical state which we call love.[11]

Love, like anger and fear, was an important emotion because it made much of the behaviour in the animal world explicable in relation to basic instincts, and so helped Spencer to make his evolutionary points at the expense of the argument from design.

An extended metaphor introduces the second part of Spencer's 'Personal Beauty': it is worth quoting in full as it perfectly illustrates the crux of the relation of character to aspect:

Imagine a book of which the first page when analysed turned out to contain a mixture of the description of *two objects nearly allied but not so identical,* expressed in ways almost alike but not quite so. Imagine that in one part of the page the sentences of the two descriptions come alternately; that in another, half-sentences from each were united into one sentence, so as to make but obscure sense; and that in some case the interchange occurred several times in the same sentence. It is clear that though you might very well recognise the nature of the things treated of, no definite conception would be conveyed to you.

Suppose further that on reading over several pages you found each of them to contain somewhat similar pairs of descriptions somewhat similarly mixed – the objects described being always akin to the first and to each other, and the manner of combining the descriptions having more or less resemblance. Possibly on comparing them you might gain some insight into the principle of arrangement, and so get a glimmering of the specific interpretation.

But now suppose that as you advanced you found *the objects treated of on the same page were in many cases more widely divergent,* and the *intermixture of the descriptions* less similar in method to foregoing ones – that beside this you by-and-by came upon pages containing a union not of two descriptions but of more, a compound of two of these compounded descriptions – and that by the time you reached the middle of the book this *jumbling of descriptions* had produced a *high degree of complexity* both in respect of the number combined and the modes of combination. What would be the result? Manifestly you would abandon all efforts at interpretation, and would doubt whether there was any meaning to be discovered. However really systematic the structure of each page, and however comprehensible to one having the clue, yet in the absence of a clue the contradictions, the inconsistencies, the mystifications would be so numerous as to suggest the suspicion that the book was an elaborate hoax [my italics].[12]

The subject-matter of these pages, Spencer explained, is typical of character whilst the description is typical of features. Then the fun starts. In the first case, the 'two objects nearly allied but not so identical' exemplify two races of man, united through joint offspring; in the second case, 'more widely divergent objects' and 'intermixture of descriptions' represent the faces produced by the intermarriage of strongly contrasted races; and in the third case, the pages with the 'high degree of complexity' and a 'jumbling of descriptions' stand for the faces of most people which develop from the repeated mixture of mixed races. Spencer's claim is that it is precisely the mixing of races with the resulting 'heterogeneity of constitution' which explains the 'incongruities between aspect and character which we daily meet with' and means that we cannot assume that mind and physiognomy are related. If, however, we take the time to consider the example of sheep breeding, he goes on to say, we will learn that the basic truth that crossing a pure breed with an impure one will produce a pure breed with a mixed constitution. He explained this as follows:

An unmixed constitution is one in which all the organs having for innumerable generations worked together, are in exact fitness, are perfectly balanced; and the system as a whole is in perfect equilibrium. A mixed constitution, on the contrary, being made up of organs belonging to two separate sets cannot have them in exact fitness, cannot have them perfectly balanced; and a system in comparatively unstable equilibrium must result. But in proportion to the stability of the equilibrium will be the power to resist disturbing forces. Hence when two constitutions in stable and unstable equilibrium respectively, become disturbing forces to each other, the unstable one will be overthrown and the stable one will assert itself unchanged.[13]

In other words, whether the actual appearance is pure or impure, it does not follow that the constitution will be unmixed or mixed accordingly. What does follow, though, is that the relationship between 'beauty of character and beauty of aspect' is provisional upon the meanings which can be given to expressions, and in particular to beautiful ones.

At issue was the nature of emotion itself: in what way could a private and personal experience, like the expression of love, assume a universal meaning? How was it expressed in women as opposed to men? Can the subjective experience of an emotion be represented? And to what extent is the term 'feeling' used to stand for a complex configuration of sensation, perception and consciousness? This chapter will consider the relationship between the physiognomical tradition and the emergence of a new evolutionary (and ultimately eugenic) fascination with healthy physiques and beauty in the light of these questions. The likes of Herbert Spencer and Alexander Bain discuss the relationship between expression, feeling, and mind in terms of the physical actions of the body, defining subjectivity as the product of progressive development whilst pondering the rationale for the regulation of emotional states. The problem was to what extent the will really was an agent in the mental processes. I will first consider the debates about appearance, character, and intellect which appeared in the middle third of the nineteenth-century. Focussing on the work of Alexander Walker and John MacVicar on beauty, I shall show the intimate relation of feeling to sensation, perception, and consciousness. I will then discuss Bain's account of the relation of mind to body, before reflecting on the nature and role of feeling in defining (female) subjectivity in the novels of Wilkie Collins. My aim in this chapter is not to survey the wealth of writings on or about the emotions[14] but to select a number of writings which, I suggest, exemplify popular thought on emotional experience, and also demonstrate the pervasive influence of physiognomic ideals.

What unites Walker, MacVicar, Collins, Bain, and Spencer is a fascination with the quality of subjective experience; the difference between them is simply the purpose their work serves. Any alignment of appearance with character deems beauty to be the highest sign of mental development; that much is affirmed in Walker's study of beauty and reworked in Collins' sensation fiction. In particular, the connection between women, beauty, and subjectivity resonates throughout nineteenth-century culture. Robyn Cooper has pointed out 'the general preoccupation of many Victorian men and a number of Victorian women with the subject of woman'.[15] 'The "speaking" of women and beauty', she explains, 'is multivocal and conflictual and the subject is . . . complicated . . . because of the connection of the female body with the emotions of love and the sensations of desire'.[16] Cooper's identification of the importance of emotion and sensation to descriptions of the female body is significant because it points to the difficulty of talking about emotion and delineating its character. In the hands of Alexander Walker, this difficulty seems to dissolve, for we are offered an analysis of beauty as an external object and a state of mind with an associated state of feeling.

II

Trained as a physiologist, Walker was, like Bell, a product of the Scottish Enlightenment tradition based in Edinburgh; his account of beauty combines 'ideas of goodness, of suitableness, of sympathy, of progressive perfection, and of mutual happiness' in order to provide a moral backdrop to the correspondence between the mental (psychical) and the physical.[17] In the advertisement to Walker's *Beauty: Illustrated Chiefly by an Analysis and Classification of Beauty in Woman* (1839) the author firmly declared not only the universal appeal of a critique of beauty in woman but also the apparent absence of such an account of beauty in nineteenth-century writing.[18] 'There is', the advertisement states, 'no subject more universally or more deeply interesting than that which is the chief subject of the present work'. 'Yet no book', the self-promotion continues, 'even pretending to science or accuracy, has hitherto appeared upon it . . . Not one has been devoted to woman, on whose physical and moral qualities the happiness of individuals and the perpetual improvement of the human race are dependent.'[19] Thus, the subject of Walker's book is woman because the happiness as well as the progress of the human race depends on this kind or type of individual. Whilst aligning Walker's work with such notable theories of beauty as the eighteenth-century accounts offered by Hume, Hogarth, Burke, Knight, and Alison, not to mention the classical accounts of Leonardo da Vinci, Winckelmann, Mengs and Bossi, the advertisement emphasises the distinctiveness of his 'new view of the theory of beauty' and its combination of 'anatomical and physiological knowledge with the critical observation of the external forms of woman'.[20] As a type, in fact, 'woman' was invested with a very particular character and assumed to involve a special kind of experience. It is, Walker subsequently explained, an account of 'the laws regulating beauty in woman, and taste respecting it in man'.[21] What both the advertisement and the text itself suggest is that the application of the scientific disciplines of anatomy and physiology to the female physique provides an opportunity to refine our rational understanding and aesthetic judgement of the beautiful. Hence, the 'mystical and delusive' character of female beauty can be replaced with an account which aimed to 'unravel the greater difficulties which that subject presents'[22] and to make beauty 'the external sign of goodness in organization and function'.[23] The implication is that it is easier to philosophise (in abstract terms) than to describe (in actual terms) the character of female beauty.

The writers alluded to in the title of Walker's *Beauty* book are exemplary of two distinct philosophical traditions: on the one hand there are eighteenth-century theorists of beauty and on the other the aesthetic practitioners of the Renaissance. Actually, Walker went to some length to align himself with the Renaissance tradition exemplified by Da Vinci in preference to the eighteenth-century tradition led by Hume.[24] Indeed, he sought to stabilise beauty by reifying it, marking out a 'new view of the theory of beauty' by reconfiguring 'woman' at the centre of the discussion. But what were the guiding ideas which motivated his project? 'In this work', he wrote, it is the form of woman which is chosen for examination, because it will be found . . . to involve knowledge of the form of man, because it is best calculated to ensure attention from men, and because it is men who, exercising selection, have alone the ability thus to ensure individual happiness and to ameliorate the species, which are the objects of this book.[25]

A few points are important in this passage. The first is that the originality of Walker's thesis seems to lie in his understanding of the relations of the sexes: 'the form of woman . . . involve[s] . . . the form of man'. The second is that these relations appear to be determined by an instinct to survive: 'men . . . exercising selection'. The final point is that this survival instinct can lead to the progress of the human race: 'men . . . have alone the ability to ensure individual happiness and to ameliorate the species'. Placed within a frame of reference that always refers to man, therefore, woman issues forth a physical knowledge that, according to Walker, presents a hermeneutic opportunity and brings with it the possibility of corruption. 'Be it known', he decreed, 'that the critical judgement and pure taste for beauty are the sole protection against low and degrading connexions'.[26] And, somewhat later, he added: 'our vague perceptions . . . and our vague expressions respecting beauty will be found to be, in a great measure, owing to the inaccuracy of our mode of examining it, and, in some measure, to the imperfect nomenclature which we possess for describing it'.[27] To put this in other words, the issue is what it means to say someone or something is beautiful. Is beauty an abstract, mental concept or an actual, physical fact? Moreover, what conception of (female) character and aspect might be involved in this sort of distinction?

True to Walker's primitive developmental thesis, there are, he announced, three kinds of beauty – locomotive, vital, and mental – and they are derived from three different classes of human organs – levers, cylindrical tubes, and nervous particles. In sum, locomotive beauty is figured by the limbs; vital beauty by the stomach; and mental beauty by the face and head (plate 16). 'It is evidently the locomotive or mechanical system whose figure is precise, striking, and brilliant', he explained:

It is evidently the nutritive or vital system which is highly developed in the beauty whose figure is soft and voluptuous. It is not less evidently the thinking or mental system which is highly developed in the beauty whose figure is characterized by intellectuality and grace.[28]

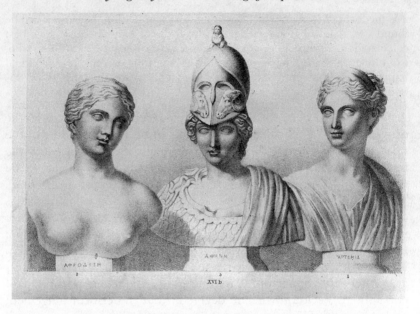

Plate 16   Alexander Walker, Three types of face, 1846.

These three kinds, which bear more than a passing resemblance to Lavater's three kinds of man – the animal associated with the belly, the moral with the breast and heart, and the intellectual with the head and eye – quite clearly gauge character from physique. What is more, important hierarchical levels of explanation can be glimpsed in Walker's schema. In reverse order, the third and lowest order is the locomotive or physical, classified as mechanical and characterised by the limbs; the second order is the nutritive, classified as vital and characterised by the stomach; and the first and highest order is the thinking, classified as mental and characterised by the face and head. But, as Walker declared, the trouble is that the 'art of distinguishing and judging of beauty in woman' confronts the informed observer with the unmistakeable truth that just one of these kinds – namely, vital beauty – prevails when describing beauty in woman. This claim is extended in the companion volume to *Beauty*, entitled *Woman*, wherein Walker recapitulated his theory as follows:

I have in my work on BEAUTY, shown that beauty of the mental or thinking system is less proper to woman than to man – is less feminine than beauty of the vital or nutritive system; and that it is not the mental, but the vital system, which is, and ought to be, most developed in woman. – Still less is it mere cerebral or intellectual, considered apart from mere sensitive beauty, which ought to characterize her.[29]

Perhaps it is not surprising, given his earlier comments, that the common type is vital beauty; that is, a beauty located in the stomach region and described characteristically as 'soft and voluptuous'. Vital beauty signals the reproductive capacity of woman, answering the question of what woman is for in terms of the growth and propagation of the species, and importantly, it is a beauty which is fitting to woman and imperatively so: 'the vital system . . . is, and ought to be, most developed in woman'. But what of the notion that superior physical beauty is the expression of superior mental development?

Walker's ideal form of (female) beauty is, we recall, the thinking kind, classified as mental and characterised by the face and head. Nowhere is this kind better imagined than in the Venus de Medici, a well-known Greek statue upon whose physique the eighteenth-century propagandist Johann Joachim Winckelmann bestowed a potent erotic charge.[30] Alex Potts explains: 'His [Winckelmann's] is a very complex reading of the formal purity of the ideal figure, in which a deathly stillness mingles with eruptions of desire and violent conflict. A powerful dialectic is set up between beautiful bodily form and suggestions of extreme psychic and physical disquiet. The figure of the Venus de Medici confronts the observer with exactly the embodiment of "woman at that age when every beauty has just been perfected".'[31] Justifying his use of the Venus as the exemplar of the first order of beauty, Walker explained the attraction with undisguised pleasure:

These exquisite details [of the head, eyes, nose, and mouth], and the omission of nothing intellectually expressive that nature presents, have led some to imagine the Venus de Medici to be a portrait. In doing so, however, they see not the profound calculation required for nearly every feature thus embodied. More strangely still, they forget the ideal character of the whole: the notion of this ideal head being too small is especially opposed to such an opinion . . . In short, I know no antique that displays such profound knowledge, both physiological and physiognomical, even in the most minutest details; and all who are capable of appreciating these things, may well smile at those who pretend to compare with this any other head of Venus, now known to us.[32]

Considering the implications of such a form makes visible the pleasures of a private knowledge, and so Walker sought to cast his desire for the Venus as a desire for a 'natural beauty'. The purpose of this move was to license an intimate knowledge of the Venus's physique in specifically moral, rather than erotic, terms. Actually, Walker overstressed this point. Ideal models of beauty were pleasing 'not merely because their forms are disposed and combined so as to effect agreeably the organ of sight, but because their exterior appears to correspond to *admirable qualities*, and to announce *an elevation* in the condition of humanity'; and, he went on, 'such do the Greek monuments appear to physiologists and philosophical artists whose minds pass rapidly from the beauty of forms to that locomotive, vital, or mental *excellence* which it compels them to suppose'.[33] The movement from 'beauty of forms' to what Walker describes as 'that locomotive, vital, or mental excellence' can only be achieved by seeing beauty as the physical, external index of the excellence (or otherwise) of the internal state.

Now, as Cooper reminds us, it is important to place Walker's work in the context of the Evangelical tradition of the time with its growing unease about representations of the female body, and especially the nude.[34] Short of explicitly promoting the appeal of 'living nudity', Walker believed the life class at the Royal Academy and the dissecting room of the London Medical Schools were suitable spaces for the display of 'natural purity', without running the risk of offending the traditional mores of climate and custom: 'the familiarity of both these classes with natural beauty leads them only to seek to inform their minds and to purify their taste'.[35] That said, it was an often-held notion that the beauty of woman was a source of illicit and possibly corrupting desire because, to paraphrase a complex argument, it displays the possibility of a pleasure that may not be openly offered to the public gaze.[36] In other words its promise of pleasure lies in the possibility of private pleasures being converted into public display. The politicised aesthetics of Herbert Marcuse, writing earlier in this century, are interesting on this theme. He claimed that 'the medium of beauty decontaminates truth and sets it apart from the present. What occurs in art occurs with no obligation.'[37] Placed in the private realm of culture, separate from the public sphere of commodified production and exchange, Marcuse makes clear that the value of beauty exists in the ideality of its form.[38] Indeed, more recently, Lynda Nead has argued that the (female) nude confronts the observer with a challenge, perhaps even a dilemma, about the nature of subjectivity:

If the female body is defined as lacking containment and issuing filth and pollution from its flattering outlines and broken surface, then the classical forms of art perform a kind of magical regulation of the female body, containing it and momentarily repairing the orifices and tears. This can, however, only be a fleeting success; the margins are dangerous and will need to be subjected to the discipline of art again . . . and again. The western tradition of the female nude is thus a kind of discourse on the subject, echoing structures of thinking across many areas of the human sciences.[39]

When that ideal form is the nude, as in Walker's case, it may well be that what is represented is not so much aspect as character. But what does it mean to be beautiful? What (if any) value is attached to the condition of 'being beautiful'? Is the category of the beautiful (and the notion of beauty) a personal, individual experience? Or, indeed, is beauty a dimension of social structure as well as an element of individual consciousness? In what ways was the beautiful defined, evoked and elaborated in physical terms via the organisation of the senses and the intellect? All these questions can, according to MacVicar and Bain, be answered once we have revised our understanding of the place of feeling in expressing our mental character.

III

Grappling with the philosophical problem of the relation of the abstract concept to the actual fact of beauty, John MacVicar made the following observation:

Suffice it to say, that, as everyone who uses the term *Beauty* in the natural exercise of his taste, understands by it a something in an external object of regard, and not a state of mind in the observer, I shall uniformly make use of the term in this sense, calling the corresponding state of feeling *the emotion of the beautiful*.[40]

A direct contemporary of Walker's, and bound by the selfsame Scottish philosophical tradition, MacVicar points out that 'beauty' is to be found in the observed *not* the observer. Yet beauty brings with it a 'corresponding state of feeling' which MacVicar calls *'the emotion of the beautiful'*. So we have an object, which is beautiful, and a response to that object, which is emotional, and these lead to the following equation: beauty is to emotion as objective is to subjective and public is to private. What then, MacVicar asks, should we make of the experience of enjoyment which occurs when we look at an object of beauty? He claims the answer lies in the type of beauty observed (object or subject) for there are two kinds of beauty, simple and expressive, and two orders of each kind:

SIMPLE BEAUTY: (divided into two orders) the members in the first order characterized, the one by hardness, angularity, and brilliancy, of the objects included in it, and of which the spectra of the *Kaleidoscope* may be taken as a type; the members in the second order by the softness, continuity, and grace of the objects included in it, and of which *Arabesques* may be taken as a type, and a Grecian vase as an example.

EXPRESSIVE BEAUTY: (divided into two orders) the one, including objects remarkable for their variety, ruggedness and spirit, but yet for a certain finiteness also, both in their extent and expression, which admits of their being easily mastered by the imagination at sight; and which coincides with what is commonly called *the Picturesque*; the other, including objects remarkable for a certain power of vastness, infinity or sameness, in their expression, before which the imagination of the beholder expands, attempting, but in vain, to embrace them; and which coincides with that which is called *the Sublime*.[41]

The object of MacVicar's enquiry was to classify the sorts of beauty and the varieties of its form so that we are not deceived into thinking that our experience of enjoyment on seeing a beautiful object is equivalent to the beautiful object itself. The only defence against such a confusion – a confusion which effectively transfers subjective impressions onto an objective form – and the only means of making sense of that transference is to recognise the order which lies beneath the beautiful, external appearance. This is how he expressed it diagrammatically:[42]

		Fitness
		Utility
	derivative from	Imitation
		Reminiscence
		Associations in general
BEAUTY		
		Ignorance
		Vicious education
		Peculiar organisation
		Flow of gaiety
	factitious by	Love
		Music
		Wine
		Anything producing a vacant flow of rapture

Something of the same desire to see order amidst confusion is evident in a comparable description of beauty given by Spencer. He wrote:

Take, as an example, that state of mind produced by seeing a beautiful statue. Primarily, this is a continuous perception – a co-ordination of the various visual impressions which the statue gives, and a consciousness of what they mean; and this is what we class a purely intellectual act. But it is impossible to perform this act without a greater or less feeling of pleasure – without some emotion. Should it be said that this emotion results from the many ideas associated with the human form; the rejoinder is, that although these may aid in producing it, it cannot be altogether so accounted for: seeing that we feel a similar pleasure on contemplating a fine building. In brief, seeing that in all cases, the materials dealt with in every cognitive process, are either sensations, or the representations of them; and seeing that these sensations, and by implication the representations of them, are always in some degree agreeable or disagreeable; it follows, of necessity, that no act of cognition can be *absolutely* free from emotion, but that the emotion accompanying it will be strong or weak, according as the materials co-ordinated in the cognition are great or small in quantity or intensity.[43]

By appropriating the mechanisms of associationist psychology, Spencer seems to have tried to extend the individual experience of learning by association to the whole race of man. The importance of associationist ideas to his understanding of feeling cannot be over-emphasised, especially those of Hartley which we glimpsed in the first chapter. In a similar fashion to Spencer, Alexander Bain, author of works on education, logic, and psychology and founder of the journal *Mind* (1876), was instrumental in bringing together associationist ideas with his own ideas on physiology and phrenology.[44] Bain elaborated and developed associationist thinking about the constitution of mind and body so that for him, all experiences, affections, and the will could be resolved into sensations and ideas. The guiding principle which directed Bain's thought was a conviction that mental and physical forces were correlated and unified: 'mind, as known to us in our own constitution, is the very last thing that we should set up as an independent power, swaying and sustaining the powers of the natural world'.[45] The idea was that the exercise of the will was dependent on the existence of physiological organisation and purely physical conditions. The result, Bain claimed, was an understanding of mind–body relations as a sort of parallel-processing system wherein the facts of consciousness (mind) are interdependent with the physical changes in the organism (body). But what of the emotions? Where does this understanding of mind–body relations get us in our discussion of the place of feeling in defining and illustrating our mental nature, the very basis of subjectivity?

This is what Bain said about the relation of mind to body:

Mind, according to my conception of it, possesses three attributes, or capacities.

1. It has Feeling, in which term I include what is commonly called Sensation and Emotion.
2. It can Act according to Feeling.
3. It can Think.

Consciousness is inseparable from the first of these capacities, but not, as appears to me, from the second or the third. True, our actions and thoughts are usually conscious, that is, are known to us by an inward perception; but the consciousness of an act is manifestly not the act, and although the assertion is less obvious, I believe that the consciousness of a thought is distinct from the thought. To flee on appearance of danger is one thing, and to be conscious that we apprehend danger is another.[46]

There are a number of proposals in this passage: mind is capable of feeling, action, and thought; feeling involves sensation and emotion; and most importantly, there is no feeling (and so no sensation or emotion) distinct from the consciousness of that feeling. Feeling is, in other words, an internal, mental state: it is the primary mark of mind and the means of distinguishing the human race from brute creation and vegetable and mineral worlds.[47] How do we know this is the case? Bain explained:

It is each in ourselves that we have the direct evidence of the conscious state, no one person's consciousness being open to another person. But finding all the outward appearances that accompany consciousness in ourselves to be present in other human beings, as well as, under some variety of degree, in the lower animals, we naturally conclude their internal state to be the same with our own. The gambols of a child, the shrinking from a blow, or a cry on account of pain, and the corresponding expressions from mental states common to all languages, prove that men in all times have been similarly affected. The terms for expressing pleasure and pain in all their various forms and degrees are names of conscious states. Joy, sorrow, misery, comfort, bliss, happiness are a few examples out of this wide vocabulary.[48]

This is a noteworthy passage as Bain is suggesting that even though the experience of feeling has a private and individual range of reference, it has, nonetheless, a public and universal sense. What he is positing, then, is an account of the subjectivity of feeling which assumes that an individual point of view can be resolved from something that only I know – for instance, my joy or my sorrow – into an affective state that everyone knows – for example, joy and sorrow.[49] This claim for the shared nature of emotional feelings does not dissolve subjective qualities of experience, but it does make them accessible to public meanings; hence, the difference between private emotional feelings and their public meanings is simply a difference of degree not kind.

Of course, this makes the task of analysis much harder. As Bain pointed out, if we want to describe states which come under the name 'feeling', and we want to make this the basis for a descriptive method, then we must look for their distinctive characteristics, 'the peculiarities, or descriptive marks, that characterize them as feelings'.[50] Take, for example, pain:

The expression of acute pain is strong and characteristic . . . The body is driven into movements and attitudes of a violent, intense character; sometimes the ordinary movements are quickened and at other times contortions and unusual gestures are displayed. The suddenness, quickness, intensity, of the bodily action, rather than the peculiar direction or form of it, constitute the distinctive character of the situation . . . If next, we turn to the *features*, whose chief use is expression, we find a much more distinctive manifestation. There is a well-known form of the countenance that marks the condition of pain, – being produced by certain movements of the eyebrows and the mouth to be afterwards analysed; and in the case of acute pains these movements have the same appearance of violence and intensity that belongs to the bodily gestures at large. The *voice*, also a medium of expression, sends out acute cries sufficient to suggest suffering to every listener.[51]

The explanation is straightforward: outward appearances, namely expressions or bodily states, correspond to internal (mental) states. Whilst the 'suddenness, quickness, [and] intensity' of bodily action are what form the descriptive marks of pain, it is clear they are most visible and legible in the expressions of the face. Therefore, 'the shrinking from a blow, or a cry on account of pain' can be proven to refer to both the subjective quality of the experience (the actual personal fear or pain) and their typical and explicitly social sense.

Under the watchful eyes of Spencer, Walker, MacVicar, and Bain, we have seen how the category of the 'beautiful' and the notion of 'beauty' might involve both a private individual experience and a public universal sense; that is to say, subjective and objective states.

The terms of the debate seem to rest on the nature of the relation-ship between the mental and the physical realms. The problem is this: on being presented with a 'beauty' we are confronted with a complex experience which places the subject and the object in an acute (and usually oppositional) relationship of difference. However, accounts of the subjectivity of feeling assume that an individual point of view can be resolved into a collective affective state. Having looked at the philosophical discussions of this problem, therefore, I want to turn now to its fictional representation in the novels of Wilkie Collins and in particular to the role of feeling in defining subjectivity in terms of women, beauty, and character. I will first explore the understanding of sensation and perception in *The Woman in White* (1860) before addressing the role of self-consciousness in *Basil* (1852) and the expression of feeling as an internal state in *No Name* (1862).

IV

Recalling his first glimpse of Laura Fairlie, the elusive and enigmatic heroine of Wilkie Collins' *The Woman in White* (1860), Walter Hart-right momentarily despairs over his failure to provide an accurate portrait of Miss Fairlie. 'How can I describe her?', Hartright laments, 'How can I separate her from my own sensations, and from all that has happened in the later time? How can I see her again as she looked when my eyes rested on her – as she should look, now, to the eyes that are about to see her in these pages?'[52] Despite his anxiety, the relation that Hartright constructs between sensation, action, and perception is instructive as an articulation of the values which might be attached to 'being beautiful'. The lengthy passage that ensues gives an evocative sketch of the complexity of the correspondence between character and appearance:

The water-colour drawing that I made of Laura Fairlie, at an after period, in the place and attitude in which I first saw her, lies on my desk while I write. I look at it, and there dawns upon me brightly, from the dark greenish-brown background of the summerhouse, a light, youthful figure, clothed in a simple muslin dress, the pattern of it formed by broad alternate stripes of delicate blue and white . . . Her hair is of so faint and pale a brown – not flaxen, and yet almost as light; not golden, and yet almost as glossy – that it nearly melts, here and there, into the shadow of the hat. It is plainly parted and drawn back over her ears, and the line of it ripples naturally as it crosses her forehead. The eyebrows are rather darker than the hair; and the eyes are of that soft, limpid, turquoise blue, so often sung by the poets, so seldom seen in real life. Lovely eyes in colour, lovely eyes in form – large and tender and quietly thoughtful – but beautiful above all things in the clear truthfulness of look that dwells in their inmost depths, and shines through all their changes of expression with the light of a purer and a better world. The charm – most gently and yet distinctly expressed – which they shed over the whole face, so covers and transforms its little natural human blemishes elsewhere, that it is difficult to estimate the relative merits and defects of the other features. (pp. 40–1)

The appeal of Hartright's portrait lies in the minuteness and the particularity of his observation of physical detail and the analogy he draws between this and Laura's character: her hair 'ripples naturally as it crosses her forehead'; and her eyes 'are of that soft, limpid, turquoise blue . . . so seldom seen in real life'. Yet, he claims, there is a difference, a lack of correspondence even, between the 'water-colour drawing' of Laura and the image of her that occupies his mind. The former, Hartright explained, is an image of 'a fair, delicate girl, in a pretty light dress, trifling with the leaves of a sketch-book, while she looks up from it with truthful innocent blue eyes – that is all the drawing can say, all, perhaps, that even the deeper reach of thought and pen can say in their language, either' (p. 41). The language of 'drawing' is set apart from the language of 'thought and pen' but the problem is that in distinguishing visual from mental and verbal forms of representation in this way, Hartright draws attention to the inadequacy of both in describing accurately the configuration of physical beauty and character which Laura embodies. Although the difference between the watercolour sketch displayed on his desk, the sensational image coveted in his mind, and the fictional narrative is explained away in temporal terms, it is clear that what causes Hartright anxiety is the absence of any commensurability between the drawing, the image, and the narrative.

Observing Laura at the time of their first encounter, Hartright later transcribes this image into a water-colour drawing, which then becomes the means of reconstructing his impressions of her beauty for the reader. The image of Laura which Hartright presents may be retrospective, but its claim to authenticity is only disrupted by Hartright's self-conscious musings on the physical features of Laura's face and, perhaps more importantly, the internal state which he imputes from these features. As the embodiment of beauty and truth, Laura poses a challenge to the powers of Hartright's interpretative method. Rather than admit defeat, he invokes what we might call the subjectivity of feeling. 'Think of her', he subsequently implores, 'as you thought of the first woman who quickened the pulses within you that the rest of her sex had no art to stir'. He went on, enthusiastically:

Let the kind, candid blue eyes meet yours, as they met mine, with the one matchless look which we both remember so well. Let her voice speak the music that you once loved best, attuned as sweetly to your ear as to mine. Let her footstep, as she comes and goes, in these pages, be like that other footstep to whose airy fall your own heart once beat time. Take her as the visionary nursling of your own fancy; and she will grow upon you, all the more clearly, as the living woman who dwells in mine. (p. 42)

In effect, the watercolour must be discarded as an inadequate form of representation. But what remains? The initial impressions have long since blurred and so Hartright is left with his recollection of those impressions. These recollections form the basis for the narrative account he gives of Laura's beauty and the seductive charms of her face, and in particular her eyes.

Regarded by Hartright as a 'poor portrait' and a 'dim mechanical drawing' (p. 41) the sketch-book figure of Laura Fairlie disappoints him because it seems, by his own admission, unable to convey the 'charm' of Laura's eyes with their capacity to radiate a 'clear truthfulness of look' over her face and convert the 'little natural human blemishes' into an ideal, almost classical, beauty. Yet, he readily admits that there may well be a linguistic, as well as a representational, problem involved in expressing the seductive charms of this beauty. 'The woman who first gives life, light, and form to our shadowy conceptions of beauty', Hartright declares, 'fills a void in our spiritual nature that has remained unknown to us till she appeared':

Sympathies that lie too deep for words, too deep almost for thoughts, are touched, at such times, by other charms than those which the senses feel and which the resources of expression can realise. The mystery which underlies the beauty of women is never raised above the reach of expression until it has claimed kindred with the deeper mystery in our own souls. Then, and then only, has it passed beyond the narrow region on which light falls, in this world, from the pencil and the pen. (pp. 41–2)

The 'sympathies that lie too deep for words, too deep almost for thoughts' seem to exhaust the range of feeling and expression and instead suggest that an appreciation of beauty depends upon the experienced character of the observer, not the observed. This is precisely the sort of transference which MacVicar identified between the 'external object of regard' and the 'state of mind in the observer'. But, as this perplexingly constructed passage indicates, the problem for Hartright is how to describe that 'indescribable something which we call beauty'. And, as we have seen, Hartright is not alone; the difficulty with the category of 'beauty' and its attendant notion of the 'beautiful' is how to communicate its experience. If we read closely Hartright's rather oddly layered metaphors, we can start to see something of what has become a familiar configuration. There are two main descriptive sequences, referring to the visible and invisible order of things, woven into the narrative: thus we have 'sympathies . . . words . . . charms . . . pencil and the pen'; and 'deep . . . deep . . . thoughts . . . touched . . . feel . . . realise . . . mystery . . . underlies'. Thus, the act of interpreting the 'beauty of woman' becomes clearly implicated in a move from the impressions of beauty (sensations) to the experience of the beautiful (perception). The intriguing thing is that the 'mystery that underlies the beauty of women' only becomes a mystery at the moment when it connects with the internal impulses of Hartright's mind (soul).

It is perhaps on similar grounds that an unsigned review published in the *Saturday Review* (25 August 1860) claimed that 'like the woman in Pope, most of Mr. Wilkie Collins' characters have no character at all'.[53] The point, surely, is that the relation of the physical to the mental has been reconceived as a relationship between actual beauty and the beautiful emotion(s) it produces. An anonymous article, 'Our Female Sensation Novelists' (1863), published in the Christian journal *Our Living Age*, provides an instructive gloss on this relationship as it is conceived in female character. The reviewer identifies 'the very language of the school [of sensation fiction]' employed in the representation of the heroine as evidence of a 'drop from the empire of reason and self-control' and towards a 'consistent appeal to the animal part of our nature'. The writer claimed: 'a whole new set of words has come into use, and they are caught up and slipped into . . . to express a certain degradation of the human into the animal or brutal, on the call of strong emotion'.[54] The difficulty for this reviewer resides in the configuration of nerves, feeling, and beauty. However, according to a recent critic, Jeanne Fahnestock, the issue is not so much a linguistic as a scientific matter; the idea is that sensation fiction marks, or maybe even articulates, a shift from a classical and idealistic mode of description to a scientific and realistic one.[55] So we could say that Hartright's encounter with Laura's beautiful face dramatises the potential correspondence of physical beauty with an essential, though abstract, concept of beauty. It is certainly true that Hartright is vividly sensible to an enigmatic and elusive space in the form of female beauty as he gathers together his thoughts on Laura *and* his 'sympathies'. Nonetheless, he is prepared to admit that amidst the seductive charms of Laura's face, there was 'another impression, which, in a shadowy way, suggested to me the idea of something wanting. At one time it seemed like something wanting in her; at another, like something wanting in myself, which hindered me from understanding her as I ought.' He explained:

The impression was always strongest, in the most contradictory manner, when she looked at me; or, in other words, when I was most conscious of the harmony and charm of her face, and yet, at the same time, most troubled by the sense of an incompleteness which it was impossible to discover. Something wanting, something wanting – where it was, and what it was, I could not say. (p. 42)

Hartright's oppositions are as follows: understanding *versus* wanting and harmony *versus* incompleteness. And the hermeneutic challenge arises not from the perception that Laura has no character but from the sensation that the character suggested by her face is not self-evident. Indeed, it is as if Laura's face is recalcitrant to interpretation. Laura is beautiful but her face poses a challenge to Hartright's hermeneutic efforts as it almost seems to exist outside the bounds of sensation and its representations. Let us think about this.

D. A. Miller's influential, and strongly Foucauldian, analysis of *The Woman in White* expands this idea of the (beautiful) body as outside meaning and focusses, in particular, on the gendered implications of such a move from private and personal to public and common. Miller claims that 'by a kind of Cartesian censorship, in which pulp-as-flesh gets equated with pulp-as-trash, the emphatic physicality of thrills in such literature allows us to hold them cheap . . . sensation is felt to occupy a site entirely outside of meaning, as though in the breathless body signification expired'.[56] The point, for Miller, is that the act of reading these sensational narratives is a double interpretive act which involves recognising the sensations of the narrators' body as well as the sensations of the reader's body. In this way, Miller maintains, sensation fiction communicates 'certain things for which our culture . . . has yet to develop another language'.[57] It is clear that Collins' 'sensations' play with an act of reading, scanning, and interpreting which sexualises the relationship between the object of beauty and the observer and, Miller argues, throws open the question of the relation of male and masculine to female and feminine. Far from being simply the expression of desire, the feeling involved in talking about beauty is made manifest through the translation of a profoundly private and personal experience into a public and general one. The sensation narrative can be felt (quite literally, and that is part of the point) to express an internal state, an 'intact privacy'; that is, an unknown and apparently unmappable interior which can produce exhilarating pulses of excitement but cannot be articulated.

Ann Cvetkovich has claimed that the expression of feeling has a political dimension which renders affect naturally liberating or reactive. Arguing for sensation as a process of reification, she explains:

Sensationalism works by virtue of the link that is constructed between the concreteness of the 'sensation-al' event and the tangibility of the 'sensational' feelings it produces. Emotionally charged representations produce bodily responses that, because they are physically felt, seem to be natural and thus to confirm the naturalness or reality of the event. The tangibility (and hence 'realness' or 'naturalness') of feeling or nervous response is invested with significance as a sign of the concreteness or reality of the representation.[58]

Drawing on the work of Marxist and feminist thinkers, Cvetkovich considers the ideological investments which influence the representation of emotional responses, especially the responses of women. In effect, she proposes a functional economy in which 'affective expression forms the basis for political action and itself constitutes a political act'.[59] As a narrative strategy, affect thus becomes the means of explaining not only 'beauty of aspect' but the personal feeling which accompanies it.[60] In most cases, this is because the female characters in Collins' fiction are presented in terms of a puzzling absence or contradiction in their faces which frustrates the expectations of the observer but can be accounted for in terms of affect. Yet, it is not exactly clear what this experience of beauty (as personal feeling) involves before it has been translated from a profoundly private and personal experience into a public and general one. The reason, I suggest, is that it is consciousness, and in particular self-consciousness, which takes centre stage in Collins' fictional narratives. The narrative in *Basil: A Story of Modern Life* – a narrative that is seen entirely 'from the inside'[61] – communicates the trouble involved in defining self-consciousness and describing its effects.

In the 'Letter of Dedication. To Charles James Ward' prefacing *Basil*, Collins outlines something of his aesthetic intention. He explains that he 'founded the main event out of which this story springs, on a fact within my own knowledge':

> I have guided it, as often as I could, where I knew by my own experience, or by experience related to me by others, that it would touch on something real and true in its progress. My idea was, that the more of the Actual I could garner up as a text to speak from, the more certain I might feel of the genuineness and value of the Ideal which was sure to spring out of it . . . By appealing to genuine sources of interest within the reader's experience, I could certainly gain his attention to begin with; but it would be only by appealing to other sources (as genuine in their way) beyond his own experience, that I could hope to fix his interest and excite his suspense, to occupy his deeper feelings, or to stir his nobler thoughts.[62]

Here, Collins' interest in the exemplary qualities of the narrative far exceeds his attention to the experience of the actual. At the beginning of *Basil*, for instance, a description of the eponymous narrator's omnibus trip which first, and fatally, brought him into contact with Margaret Sherwin is tantalisingly built up as a narrative of 'woman' unveiling before watchful (male) eyes.[63] Transfixed, Basil describes Margaret's physique, and in particular her face, in an enchanting and detailed fashion:

> Her veil was down when I first saw her. Her features and her expression were but indistinctly visible to me. I could just vaguely perceive that she was young and beautiful; but, beyond this, though I might imagine much, I could see little . . . We had been moving onward for some little time, when the girl's companion addressed an observation to her. She heard it imperfectly, and lifted her veil while it was being repeated. How painfully my heart beat! I could almost hear it – as her face was, for the first time, freely and fairly disclosed! She was dark. Her hair, eyes, and complexion were darker than usual in English women. The form, the look altogether, of her face, coupled with what I could see of her figure, made me guess her age to be about twenty. There was the appearance of maturity already in the shape of her features; but their expression still remained girlish, unformed, unsettled. The fire in her large dark eyes, when she spoke, was latent. Their languor, when she was silent – that voluptuous languor of

black eyes – was still fugitive and unsteady. The smile about her full lips . . . struggled to be eloquent, yet dared not. Among women, there always seems something left incomplete – a moral creation to be superinduced on the physical – which love alone can develop, and which maternity perfects still further, when developed. I thought, as I looked on her, how the passing colour would fix itself brilliantly on her round, olive cheek; how the expression that still hesitated to declare itself, would speak out at last, would shine forth in the full luxury of its beauty, when she heard the first words, received the first kiss, from the man she loved! (pp. 29–31)

The playful movement between the veil and the face, or in effect between concealment and revelation, disguise and display, drama-tises a future seduction by imagining and making sense of what cannot be seen. For, indeed, Margaret's face and body cannot be read at this moment – they are impenetrable to Basil's eager gaze. In a phrase that finds an echo in Hartright's perception of 'something wanting' in Laura Fairlie, Basil announces that 'among women, there always seems something left incomplete'. The absence may be alluring but, as Basil eventually discovers, the challenge is how to account for that incompleteness, that 'something wanting', as the expression of an internal state of feeling and not only that. Implied in both Hartright's and Basil's descriptions of (beautiful) women is a conundrum which replays Spencer's and MacVicar's distinctions between the abstract, mental concept of beauty and its actual, physical form.

In sum, Basil's misfortune lies in his failure to comprehend the significance for his own self of the veiled and potential pleasures promised by Margaret's beautiful face. Unable to read this screen as anything other than an enthralling prologue to a relationship, Basil seems to surrender the rational powers of his mind in favour of a decidedly irrational absorption into his own nervous sensibilities.[64] Hence, the narrative lurches from the passions of his heart to the reflexes of his nerves as Basil constructs the story of his insecurity, his melancholy, and his fragile masculinity. 'In the ravelled skein', he philosophises, 'the slightest threads are the hardest to follow'. And, in a passage with remarkable overtones of Hartley, he goes on:

In analysing the associations and sympathies which regulate the play of our passions, the simplest and homeliest are the last we detect. It is only when the shock comes and the mind recoils before it . . . that we really discern what trifles in the outer world our noblest mental pleasures, or our severest mental pains, have made a part of ourselves – atoms which the whirlpool has drawn into its vortex, as greedily and surely as the largest mass. (pp. 144–5)

It is, in other words, that which is most familiar which has the greatest potential to shock: namely, the subjective experience of our own self. Through the principle of association, we can learn something of the way in which fragmentary states of consciousness arouse other fragmentary states. For our 'noblest mental pleasures' and 'severest mental pains' absorb 'trifles in the outer world' and amplify them against the pressures of the passions. The result is a precarious vulnerability which emerges from the reflexes of the nerves rather than the beatings of the heart; and the physiological resonances are all too obvious.

Despite his astute interpretation of Mrs Sherwin, Basil fails to translate the tragic signs displayed on her face into indications of the tragedy in which he is already, irrevocably, immersed:

Her pale, sickly, moist-looking skin; her large, mild, watery, light-blue eyes; the restless timidity of her expression; the mixture of useless hesitation and involuntary rapidity in every one of her actions – all furnished the same significant betrayal of a life of incessant fear and restraint; of a disposition full of modest generosities and meek sympathies, which had been crushed down past rousing to self-assertion, past ever seeing the light. There, in that mild, wan face of hers – in those painful startings and hurryings when she moved; in that tremulous, faint utterance when she spoke – there, I could see one of those ghastly heart-tragedies laid open before me, which are acted and re-acted, scene by scene, and year by year, in the secret theatre of the home; tragedies which are ever shadowed by the slow falling of the black curtain that drops lower and lower every day – that drops, to hide all at last, from the hand of death. (pp. 75–7)

Even the conclusion to Basil's narrative expresses a sense of profound discontent with the need to step out of this form of sensational affinity in order to end the story. In a painfully personal passage, he implores the reader:

How are the pages which I am about to send to you to be concluded? In the novel-reading sense of the word, my story has no real conclusion. The repose that comes to all of us after trouble . . . is the end which must close this autobiography: an end, calm, natural, and uneventful; yet not, perhaps, devoid of all lesson and value. Is it fit that I should set myself, for the sake of effect, to make a conclusion, and terminate by fiction what has begun, and thus far, has proceeded in truth? In the interests of Art, as well as in the interests of Reality, surely not! (p. 339)

The suggestion, quite simply, is that the conclusion of the narrative marks the conclusion of Basil's life in so far as the sensational body that makes up the narrative has, apparently and fatally, come to the end of its material existence. It might well be that the lesson of sensation fiction is a denial of its very fictiveness and an assertion of the concepts of value and truth over affect. But, to end the matter here would be to ignore the function of the (exclusively) male narrators in responding to the variously beautiful female faces which confront them. What remains incompletely explained is the extent to which the male narrator is complicit in the expression of feeling. Or, to put it slightly differently, is the expression of something wanting actually the articulation of an inability to read the expressions of the face? To think about this, I want to consider the first meeting between Basil and Mannion; an encounter that is notable both for its description of a male countenance (Mannion's) and its dramatisation of the relationship between beauty, character, and expression.

Deeply, and at this time happily, immersed within the intrigue of his secret marriage to Margaret, Basil places his meeting with Mannion within the context of a series of changes which began to unsettle 'the calm uniformity of the life at North Villa'. The flurry of restless activity that anticipates the expected arrival prompts Basil to ask, impatiently, 'who was this Mr. Mannion, that his arrival at his employer's house should make a sensation?' His question is soon answered when Mannion enters the room but, as he quickly discovers, there is something in the physical appearance of Mannion that leaves his desire for information unsatisfied. 'I looked at him with a curiosity and interest', Basil recalls, 'which I could hardly account for at first': (p. 109)

Viewed separately from the head (which was rather large, both in front and behind) his face exhibited, throughout, an almost perfect symmetry of proportion. His bald forehead was smooth and massive as marble; his high brow and thin eyelids had the firmness and immobility of marble, and seemed as cold; his delicately-formed lips, when he was not speaking, closed habitually, as changelessly still as if no breath of life ever passed them. There was not a wrinkle or line anywhere on his face. (p. 110)

The reference to classical form is clear in the terms of description: here is a man who appears, to Basil, to be more like a statue than a living person. At first, physique and character are intimately related, and then they are denied; thus, the parallel with the classical ideal is sustained in Basil's analysis of the expressiveness of Mannion's face. 'Such was his countenance in point of form', he says, 'but in that which is the outward assertion of our immortality – in expression – it was, as I now beheld it, an utter void':

Never before had I seen any human face which baffled all inquiry like his. No mask could have been made expressionless enough to resemble it; and yet it looked like a mask. It told you nothing of his thoughts, when he spoke: nothing of his disposition, when he was silent. His cold grey eyes gave you no help in trying to study him. They never varied from the steady, straightforward look, which was exactly the same for Margaret as it was for me; for Mrs Sherwin as for Mr Sherwin – exactly the same whether he spoke or whether he listened; whether he talked of indifferent, or of important matters. Who was he? What was he? His name and calling were poor replies to those questions. Was he naturally cold and unimpressible at heart? or had some fierce passion, some terrible sorrow, ravaged the life within him, and left it dead for ever after? Impossible to conjecture! There was the impenetrable face before you, wholly inexpressive – so inexpressive that it did not even look vacant – a mystery for your eyes and your mind to dwell on – hiding something; but whether vice or virtue you could not tell. (pp. 110–11)

Fuelled by a sense of the impenetrability of Mannion's face, the description betrays an anxiety as to what this marble face might signify, which is in turn translated into a nervousness about the powers of perception and comprehension. For instance, Mannion's face is on the one hand, 'an utter void' and too 'expressionless' even for a mask; and on the other hand it is suggestive of meaning 'like a mask', and 'hiding something'. In effect, Mannion is 'a complete walking mystery' (p. 114) to Basil, and it is his very lack of expression, his expressionlessness, which dramatises the problems associated with the expression of emotion. In this sense, Basil's failure to read and interpret Mannion's character is as much a problem of feeling as a misreading of the correspondence between character and aspect.

There is nothing more obvious to each of us than that we are the subject of experience; we delight in sensations and perceptions, we suffer grief or pain, we have ideas and we are aware that we can think. And yet, there is nothing more difficult to us than to describe and explain what this consciousness is and what form it takes. It may be that individually we are conscious of the ways in which we are conscious of things, as this determines the experience of being ourselves. But when we consider what others are conscious of and what form their consciousness takes 'from the outside', we have to rely on various observable features as signs of that which each conscious subject knows 'from the inside', as it were. Under what conditions does sensation occur? Are there any limitations to its experience? What does it mean to be sensational? Is it an emotional state and, if so, can this state be distinguished from that of being beautiful and does it bring with it any special responsibilities? All these questions come down to the subjectivity (or consciousness) of feeling. The problem is that consciousness is not transparent to itself but is hidden from view; it is named 'character' as it makes us who we are but it is discernible only through sensation.

In Collins' *No Name*, Miss Garth's reflections on hearing of the dreadful ignominy of illegitimacy facing her former pupils, Norah and Magdalen Vanstone, constitute a now familiar search for a convincing answer to the nature of character and consciousness:

Does there exist in every human being, beneath that outward and visible character which is shaped into form by the social influences surrounding us, an inward, invisible disposition, which is part of ourselves; which education may indirectly modify, but can never hope to change? Is the philosophy which denies this, and asserts that we are born with dispositions like blank sheets of paper, a philosophy which has failed to remark that we are not born with blank faces – a philosophy which has never compared together two infants of a few days old, and has never observed that those infants are not born with blank tempers for mothers and nurses to fill up at will? Are there, infinitely varying with each individual, inbred forces of Good and Evil in all of us, deep down below the reach of mortal encouragement and mortal repression – hidden Good and hidden Evil, both alike at the mercy of the liberating opportunity and the sufficient temptation? Within these earthly limits, is earthly Circumstance ever the key; and can no human vigilance warn us beforehand of the forces imprisoned in ourselves which that key may unlock?[65]

Basing her inquiry on the anti-Lockean hypothesis that 'we are not born with blank faces', Miss Garth's philosophical debate on mind and morals is engrossing for its advocacy of psychological determinism.[66] The suggestion is, of course, an essentialist one which affirms the physiognomic method of interpretation, and yet it is implied that there is a risk in this kind of determinism insofar as it may not be able to account for the internal state of feeling. Thus we are introduced to Magdalen Vanstone through her physical appearance. She 'presented no recognizable resemblance to either of her parents', and the narrator asks, 'how had she come by her hair? how had she come by her eyes?' As if to work this out, the narrator starts to sketch Magdalen's face in detail:

Her eyebrows and eyelashes were just a shade darker than her hair, and seemed made expressly for those violet-blue eyes, which assert their most irresistible charm when associated with a fair complexion. But it was here exactly that the promise of her face failed of performance in the most startling manner. The eyes, which should have been dark, were incomprehensibly and discordantly light: they were of that nearly colourless grey, which, though little attractive in itself, possesses the rare compensating merit of interpreting the finest gradations of thought, the gentlest changes of feeling, the deepest trouble of passion, with a subtle transparency of expression which no darker eyes can rival.[67]

It is Magdalen's eyes which possess 'a subtle transparency of expression' and as such, articulate 'the finest gradations of thought, the gentlest changes of feeling, the deepest trouble of passion'; the implication is that her character can be understood in precisely these terms.

A skilful (and ironic) review of the character and significance of Magdalen Vanstone, written, though unsigned, by Margaret Oliphant, takes up this question of the moral character which may (or may not) lie behind her physique. She writes:

Mr Wilkie Collins, after the skilful and startling complications of the Woman in White – his grand effort – has chosen, by way of making his heroine piquant and interesting in his next attempt, to throw her into a career of vulgar and aimless trickery and wickedness, with which it is impossible to have a shadow of sympathy, but from all the pollutions of which he intends us to believe that she emerges, at the cheap cost of a fever, as pure, as high-minded, and as spotless as the most dazzling white of heroines. The Magdalen of *No Name* does not go astray after the usual fashion of erring maidens in romance. Her pollution is decorous, and justified by law; and after all her endless deceptions and horrible marriage, it seems quite right to the author that she should be restored to society, and have a good husband and a happy home.[68]

Placed within a frame of reference which foregrounds such models of female character as 'the most dazzling white of heroines' and 'maidens of romance', it is, perhaps, not surprising that Magdalen Vanstone's actions are judged wanting. However, what causes Oliphant the greatest alarm, apparently, is the universal nature of the appeal that sensation fiction makes to 'the passions and emotions of life'.[69] These narratives, Oliphant claimed, rely on a 'simple physical effect' which is 'totally independent of character, and involves no particular issue . . . The effect is pure sensation, neither more nor less; and so much reticence, reserve, and delicacy is in the means employed . . . that the reader feels his own sensibilities flattered by the impression made upon him.'[70] And a contemporaneous review of female sensation novelists makes clear that this 'appeal to the nerves rather than to the heart' serves only to convey 'a picture of life which shall make reality insipid and the routine of ordinary existence intolerable to the imagination'.[71] The contrast of these views with the PRB and their art of the ordinary could not be greater. At the very least, the idea is that sensation goes beyond the bounds of 'ordinary existence' and in so doing, it draws attention to, sometimes even uncovers, the nervous system which lies at the centre of our physical and emotional life.

V

As the quotation at the head of the chapter indicates, Bain proposed that the mind was made up of an inward feeling (or consciousness) and an outward action (or expression). To Bain, the mind and body had to be integrated in this way as the purpose of action or expression was to make inward feeling visible to others 'from the outside'. Thus far, we have seen the ways in which feeling can be used both to express sensation and consciousness and to explain character and aspect. An analysis of feeling which aligns the 'mental' and the 'physical' in the form of consciousness and character goes a long way to answering questions about the nature and role of subjective experience. Central to this experience is the personal and private character of consciousness, and as Spencer, in particular, believed, 'personal beauty' was the sign of this internal state because 'mental and facial perfection are fundamentally connected'. An entry in *The Phrenological Miscellany: or, the Annuals of Phrenology and Physiognomy, from 1865 to 1873* provides a further and instructive critique of 'personal beauty'. 'There is nothing more attractive and fascinating', Reverend W. T. Clarke asserted, 'than personal beauty',

All men instinctively admire a handsome form and face. They go to the opera, the theatre, the church, wherever people congregate, to feast their eyes upon human beauty. They pay the highest price for the painted counterfeit of it, however imaginary the semblance to adorn their parlour walls. We do not wonder that men are so fascinated by it, and sometimes are so smitten by the sight of it, that they pine away in misery if they can not call its possessor theirs. We do not wonder that people resort to all devices and expedients to preserve and cultivate it, and that the aid of costly clothing, paints, and cosmetics are invoked to conjure up its semblance and prolong its spells.[72]

Although the issue here is, clearly, the value of beauty as a commodity, exhibited, copied, bought, and preserved, the writer nonetheless tries hard to insist that the attractions of physical beauty must refer to the state of the mind (soul). Devoting two long sections to discussions of 'How to be Beautiful' and 'How not to be Beautiful', Clarke speculates on the connection between beauty and nobility, readily acknowledging that confronted with 'a beautiful person – mankind has always gone down on its knees before it as the shrine of a god', because 'to be beautiful is one of the spontaneous ambitions of the human heart'.[73] What is more, Clarke reminds women everywhere, 'it is not only right, but a duty, to try to be beautiful'. The practical question of how to fulfil this responsibility is easily answered, for whilst 'beauty of form and feature, a particular cut, contour, and colour of face and countenance' are 'admirable', there is a higher order of beauty – 'a beauty of expression which enfolds the features in an atmosphere of indelible fascination' and 'a beauty of mind, of disposition, of soul, which makes us forget the absence of regular features and lovely tints where they are not, and overlook their presence where they are'. He continues in expansive mood: 'everybody has seen men and women of irregular features and ungraceful form who, notwithstanding their physical defects, were so irradiated and glorified by the outshining of noble thoughts and kind affections that they seemed supremely beautiful'.[74]

The determining idea, then, is that by making beauty reside in the mind (soul) rather than the body – in 'noble endeavours and holy living' rather than 'the symmetric form and the finely chiseled face' – beauty is adapted from the ideal, public form upheld by Walker and MacVicar into a real, personal activity. According to Clarke, beauty is a state of mind, an intellectual endeavour, which is stimulated through the pursuit of knowledge; it 'realizes our ideal and wins the admiration of all cultured minds'.[75] Couched within the terms of a neo-Platonic discussion on beauty, what makes Clarke's observations so noteworthy is the provision of a culture of beauty that relies on organic metaphors of growth and decay for its articulation. Beauty is continually 'built up', cultivated, fed, and replenished, and in one gloriously romantic sentence, it 'increases with age, and, like the luscious peach, covered with the delicate plush of purple and gold which comes with autumn's ripeness, is never so beautiful as when waiting to be plucked by the gatherer's hand'.[76] Translated into a description of female beauty (and with it subjectivity), this phrase becomes more than suggestively erotic but, despite his opening lines, Clarke was at pains to point out that his desire was to render our response to beautiful forms primarily an educative, rather than necessarily a voyeuristic, experience. Indeed, Clarke seems to have been committed to writing an understanding of morality into any appreciation of the beautiful.

CHAPTER 5

Universal expressions: Darwin and the naturalisation of emotion

Fear, when strong, expresses itself in cries, in efforts to hide or escape, in palpitations and tremblings; and these are just the manifestations that would accompany an actual experience of the evil feared. The destructive passions are shown in a general tension of the muscular system, in gnashing of the teeth and protusion of the claws, in dilated eyes and nostrils, in growls; and these are the weaker forms of the actions that accompany the killing of prey.

Herbert Spencer[1]

I

Charles Darwin's investigation into expression and emotion in man and animals began in his notebooks of 1838 and gained momentum in the following two years, at a time when he was already speculating on a primitive version of what would become the principle of evolution by means of natural selection.[2] The series of notebook speculations on metaphysics, morals, and expression which eventually became *The Expression of the Emotions in Man and Animals* (1872) were originally intended to form part of *The Descent of Man, and Selection in relation to Sex* (1871)[3] but a concern about the sheer quantity of material he was presenting led Darwin to publish his enquiry into expression as a separate and subsequent volume.[4] Hence, having finished the final revisions to *The Descent of Man* on 15 January 1871, he started work on *The Expression of the Emotions* a couple of days later, and it was published on 8 November 1872. Much has been written about *The Descent of Man*, which is often thought of as the second half of *On the Origin of Species by Means of Natural Selection*,[5] and as such a demonstration of the evolution of man.[6] Yet less attention has been paid to *The Expression of the Emotions*, despite the fact that it was an immediate best-seller when first published in 1872.[7] There are many possible explanations for the popularity of this work at the time, from the assimilation of Darwinian thought into late nineteenth-century culture and the accessibility of scientific knowledge (through the efforts of figures like Thomas Henry Huxley and Herbert Spencer) to the personal and anecdotal tone of the book. Explaining that by the publication of *The Descent of Man*, 'Darwinism . . . was well on its way to becoming the new orthodoxy in Victorian science and society', Evelleen Richards has reflected on the transformation of the 'preindustrial modes of legitimation' in the light of 'a secular, naturalistic redefinition of the world'.[8] The study of expression participates in this redefinition, as I have shown throughout this book, and Darwin, in particular, sought a naturalistic explanation of emotion.

What are expressions for? What is the relationship between emotions and expressions? Why do we express our feelings through bodily actions? What is the purpose of their associated mental and physical states? Are expressions habitual and instinctive, and if so, how have they been acquired? Is an understanding of expressions learned or is such knowledge innate? These are the questions Darwin asked in respect of emotional expression in men and animals. He signalled his intention to discuss the expression of different emotions in the introduction to *The Descent of Man*, prompted, he admitted, by his interest in Charles Bell's work on expression, later recalling how impressed he had been by Bell's *Anatomy and Philosophy of Expression* (1844) and the role it played in prompting him to think about the need for a rational explanation of expression.[9] As he writes in *The Expression of the Emotions*,

It seemed probable that the habit of expressing our feelings by certain movements, though rendered innate, had been in some manner gradually acquired. But to discover how such habits had been acquired was perplexing in no small degree. The whole subject had to be viewed under a new aspect, and each expression demanded a rational explanation. This belief led me to attempt the present work, however imperfectly it may have been executed. (p. 19)

To Darwin, a rational explanation of expression necessarily in-volved an illustration of the function of emotion conceived in terms of habitual movements. The challenge was, therefore, to provide an interpretation of emotional responses which emphasised the evolu-tionary purpose of the instinct to express emotions, and proved the existence of this instinct in animals as well as men. His aim was, according to Janet Browne, 'to concentrate on the evolution of expressions themselves, not the psychology of their identification; on the physical attributes, rather than mental perception and conventions'. In an intelligent essay on the subject, Browne stresses the importance of the physical mechanisms of emotional expression in mapping out the principle of continuity: 'Darwin was more interested in the way man's body actually worked, than in the theory of perception', she wrote, 'real phenomena were more useful in the fight to establish continuity between human and other species'.[10]

Central to Darwin's study of expression was a rejection of Bell's explanation of the meaning of the expression of emotions. Bell's work on expression involved three related premises: a belief in the natural association of certain emotions with specific expressions and muscles; the identification of specifically 'human' muscles for specific expressions such as blushing; and an understanding of man as a separate creation from all other animals. In sum, Bell claimed expressions had a transcendental purpose in that their origin was in instinctive actions intended to convey a theological framework: 'grimaces and smiles, frowns and blushes had been inscribed in human physiognomy', Robert Richards writes, 'as a kind of natural language which allowed immediate communication of one soul with another'.[11] The second and third of Bell's premises caused the problem for Darwin because even though he believed in the instinctive action of emotions, and also that animals had emotional responses, finding the appropriate evidence through analogy was complex. Acknowledging the problem, he wrote in *The Expression of the Emotions*:

Besides, judging as well as we can by our reason, without the aid of any rules, which of two or more explanations is the most satisfactory, or are quite unsatisfactory, I see only one way of testing our conclusions. This is to observe whether the same principle by which one expression can, as it appears, be explained, is applicable in other allied cases; and especially, whether the same general principles can be applied with satisfactory results, both to man and the lower animals . . . The difficulty of judging of the truth of any theoretical explanation, and of testing it by some distinct line of investigation, is the great drawback to that interest which the study seems well fitted to excite. (p. 18)

His interest was in telling us as much as possible about the expression being analysed, both in terms of its current state and its evolution, and in so doing he denied the transcendentalism of expression. Bell's argument for expression was, of course, based on physiognomy; thus, a rejection of the design-led argument for expression went hand-in-hand with a denial of the physiognomic foundations of expression.[12] Darwin said he sought an 'understanding [of] the cause or origin of the several expressions, and of judging whether any theoretical explanation is trustworthy' (p. 25), as he explained in detail:

Whatever amount of truth the so-called science of physiognomy may contain, appears to depend . . . on different persons bringing into frequent use different facial muscles, according to their dispositions; the development of these muscles being perhaps thus increased, and the lines or furrows on the face, due to their habitual contraction, being thus rendered deeper and more conspicuous. The free expression by outward signs of an emotion intensifies it. On the other hand, the repression, as far as this is possible, of all outward signs of an emotion softens our emotions. (p. 366)

According to him, Bell 'may with justice be said, not only to have laid the foundations of the subject as a branch of science, but also to have built up a noble structure':

It is generally admitted that his service consists chiefly in having shown the intimate relation which exists between the movements of expression and those of respiration. One of the most important points, small as it may at first appear, is that the muscles round the eyes are involuntarily contracted during violent expiratory efforts, in order to protect these delicate organs from the pressure of the blood. (p. 2)

Despite his scepticism for physiognomy, then, Darwin admired Bell's work precisely because it demonstrated the importance of the respiratory muscles to our capacity to express emotion and so moved the study of expression towards physiological grounds.

My purpose in this chapter is to demonstrate the meaning of expression in Darwin's study of the subject. I claimed at the beginning of this study that the emergence of physiological explanations for the expression of emotions constitutes a critical turn away from physiognomic ideas of the meaning of expression and towards a natural scientific conception of emotion. The evolutionary account of expression presented by Darwin is, of course, indicative of this turn as it recapitulates (and amplifies) his evolutionary ideas and synthesises them with the physiological thought of William Carpenter and Thomas Laycock, and the physiological psychology of

Alexander Bain and Herbert Spencer, in particular. Many of the basic Darwinian themes are present in *The Expression of the Emotions* – dissent from anthropocentrism, reconstruction of the circumstances which made man, and the principle of the continuity of the species – and together they construct a vision of the world in which the diversity of individual organisms produces a variety of potential futures for the species. Hence, variation is the name of the game in evolutionary theory because it helps us understand the products of evolution as well as the process by which they came to be and, importantly, it is played out at the level of the species rather than the individual. 'Darwin's species were discretely bounded groups that existed in nature, whether nature was considered in a local or a more general context', Peter Stevens has advised in a recent series of essays on leading concepts in evolutionary biology; moreover, 'Darwin was very much an exception in denying reality to the rank of species and in attaching little importance to the inability to hybridize as a distinction of rank'.[13] The problem is that evolutionary theory may not provide us with the distinctions necessary to comprehend the products of evolution as well as the process through which they came to be. To put it slightly differently, there is a question as to how far the principle of universality can be made explicable through the genealogy of species. Darwin's answer in respect of expressing emotion was simple – emotions were not useful in any way or form. Having rejected Bell's theological framework for understanding instinctive expressions via transcendental purposiveness, Darwin failed to recognise their biological function in aiding the instinct to communicate in both humans and animals in the organic world. To understand Darwin's error in perceiving emotion as useless, it is necessary to trace the background to his study of expression by looking at his early thoughts and speculations from the notebooks, as it is here that the ramifications of Lavater's teachings on the instinctive nature of physiognomic judgements become clear.

II

Darwin based his studies of expression on the observation of his own children – particularly his eldest child William Erasmus, from his birthday, 27 December 1839, to September 1844 – recording the stages of infant development, instinctive and acquired, with dispassionate objectivity and unrelenting zeal.[14] Showing no apparent sentimental attachment to his offspring, and clearly obsessed with recording the physiological elements of expression, Darwin's first entry reads as follows:

I W. Erasmus. Darwin born. Dec. 27th. 1839. – During first week. yawned, streatched [sic] himself just like old person – chiefly upper extremities – hiccupped – sneezes sucked, Surface of warm hand placed to face, seemed immediately to give wish of sucking, either instinctive or associated knowledge of warm smooth surface of bosom. – cried & squalled, but no tears – touching sole of foot with spill of paper, (when exactly one week old), it jerked it away very suddenly & curled its toes like person tickled, evident subject to tickling – I think also body under arms. – more sensitive than other parts of surface – What can be origin of movement from tickling; neck, & armpit between the toes are places seldom touched but are easily tickled – the whole surface of the sole of foot is toouched [sic] constantly. – so is the resting place of body but the latter is by no means sensitive to tickling – nor are ends of fingers, or surface of limbs – but back bone is.[15]

The attraction of observing infants, Darwin said, is that, like people of old age, 'younger children . . . look at people with a degree of fixedness which always strikes me as odd'. He went on to specify the appeal of this oddity: 'it is very like the manner older people only look at inanimate objects – I believe it is, because there is no trace of consciousness in very young children – they do not think, whether the person, they are looking at, is thinking of them'.[16] The implication is that those of young and old age look without thinking, observing objects placed before them without registering awareness. The desire to look in the young and the old is, in other words, arbitrary and unmotivated; it is unconscious and unconstrued, without reference to the physical eye or the mental 'I' of self.[17]

These journal entries make for fascinating reading in that they reveal the prominent ideas and speculations which underpin Darwin's early intellectual development.[18] What Darwin sought at this early stage of the development of his ideas was as much information as possible (and as factual as possible) about the varieties of human behavioural acts. He wanted to know the conditions sufficient to cause a visible change in the actions of an individual, especially as manifested through gestures and expression and, as the above quotation reveals, decisive in this was the difference of response between voluntary and involuntary actions. An entry in 'Notebook M' (1838) supports this view:

What is Emotion analysis of expression of desire – is there not protrusion of chin, like bulls & horses. – 1838 good instance of useless muscular tricks accompanying emotion. – when horses fighting, they put down ears, when <<turning round to kick>> kicking they do the same. although it is then quite useless – . . . Why does any great mental affection make body tremble. Why much laughter tears. – & shaking body. – Are those parts of body, as heart, & chest (sobbing) which are most under great sympathetic nerve most subject to habit, as being less so will. – . . . The whole argument of expression more than any other point of structure takes its value. from its connexion with mind, (to show hiatus in mind not saltus between man & Brutes) no one can doubt this connexion. – look at faces of people in different trades &c &c &c . . . Hyaena pisses from fear so does man. – & so dog [author's emphasis].[19]

This notebook, described by Darwin on the front cover as 'full of Metaphysics on Morals & Speculations on Expression' was written from *c.* 15 July to 2 October 1838 and was followed immediately by 'Notebook N' on 'Metaphysics and Expression' from 2 October 1838 to *c.* 20 July 1839, with the last entry made on 28 April 1840.[20] The final sentence of these notes – 'the whole argument of expression more than any other point of structure takes its value. from its connexion with mind, (*to show hiatus in mind not saltus between man & Brutes*) [my italics]' – contains the real significance as it indicates that the separation of humans from animals exists only through a break or gap (rather than a sudden transition) in the continuous development of the species. Actually, he implies that though there are specific differences between humans and animals in expressing emotions, these differences are determined not by the usefulness of the instinctive expression of emotion but by the influence of habitual actions. Hence, the difference here between the fighting horse with flattened ears and the trembling body of a fearful man is simply a difference of degree not kind, and as such is suggestive of the diversity of the species.[21]

It is easy to read all sorts of significance into these notes (and diary entries) in retrospect, but whilst they provide us with intriguing hints about Darwin's ideas and the development of his scientific project, the mode and manner of their presentation means that it is not often that we can expand these hints and fragments into clear statements of position. What we can glimpse from these speculations on emotion, though, is Darwin's persistent reasoning by analogy; the reference to Lavater in the same notebook strikingly illustrates this:

Lavater's *Essays on Physiognomy* translated by Holcroft. Vol I. p.86. 'We ought never to forget – ; that every man is born with a portion of phsiognominical [sic] sensation, as certainly as every man who is not deformed is born with two eyes.' I think this cannot be disputed anymore in men than in animals. –[22]

The capacity of animals to express emotion was evident, Darwin was sure, because its resemblance to a human expression indicated the same emotion. With a nod to Lavater, he exploited this anthropomorphic sense of continuity, applying the 'phsiognominical [sic] sensation' to men and animals as a means of showing the essential universality of many of the expressions. By this stage of his speculations, Howard Gruber noted, Darwin was 'well into the subject of expression, continuing to collect observations, reading Lavater's lengthy treatise, and, most important of all, beginning to formulate the three principles of expression with which he later will begin his book *The Expression of the Emotions in Man and Animals*'.[23] The interesting thing here is the influence of Lavater on his study of expression, to the extent that he uses Lavater's idea of physiognomic judgements as instinctive acts to demonstrate the analogous emotional expressions of humans and animals.

Earlier discussions of the common context for the consideration of the relationship of mind and body, highlighted in the first two chapters of this study, focussed in particular on the distinctions conferred on life and mind as a result of the existence of the soul, including the establishment of a barrier between humans and other organisms. It was in this context, I noted, that physiological principles, such as that of the reflex action, the nerve function, and the notion of the *sensorium commune*, emerged to support an understanding of the command of the nervous system over the individual instead of the individual controlled by mind. My point was that descriptions of human nature, of the kind given by Lavater for example, participated sometimes directly but often indirectly in complex arguments about the actions of the nerves and the muscles, the efficacy of the will, and the purpose of sensations and emotions. It is, of course, the last of these – namely the purpose of emotions – which most concerned Lavater, who insisted, like Bell (but not like Darwin), on the transcendental function of instincts expressing emotion within a theological world view. As Richards writes:

When assessing the work of natural theologians Darwin was always ready to acknowledge the functional, though not the theological, significance they assigned to animal traits. Bell did not recognize any singularly biological function of the emotions, only their transcendental purpose. When Darwin rejected the latter, he had no better idea than Bell about the former. Neither Bell nor he could imagine any strictly biological use human emotions might have. Both men were simply unaware of the kind of important communicative functions the emotions serve in the animal and human economy.[24]

An evolutionary account of expression was not concerned with transcendental explanations of physical attributes; rather, it was intended to show the development of the instincts for expressing emotions. Hence, Darwin's enquiry was directed to showing that primate expressions were explicable by the same universal principles and, therefore, man was not a separate, divinely created species.

In order to do this, Darwin tried to widen the scope of his observations from his own children to a broader sample of individuals, and also provide some sort of verification of their actuality. He circulated a questionnaire in 1867 to a number of different observers, 'several of them missionaries or protectors of the aborigines', with a request they respond with real evidence rather than just recall memories of their experience.[25] The long list of questions (with instructions as to the specific information he wanted) was as follows:

(1) Is astonishment expressed by the eyes and mouth being opened wide, and by the eyebrows being raised?

(2) Does shame excite a blush when the colour of the skin allows it to be visible? and especially how low down the body does the blush extend?

(3) When a man is indignant or defiant does he frown, hold his body and head erect, square his shoulders and clench his fists?

(4) When considering deeply on any subject, or trying to understand any puzzle, does he frown, or wrinkle the skin beneath the lower eyelids?

(5) When in low spirits, are the corners of the mouth depressed, and the inner corner of the eyebrows raised by that muscle which the French call the 'Grief muscle'? The eyebrow in this state becomes slightly oblique, with a little swelling at the inner end; and the forehead is transversely wrinkled in the middle part, but not across the whole breadth, as when the eyebrows are raised in surprise.

(6) When in good spirits do the eyes sparkle, with the skin a little wrinkled round and under them, and with the mouth a little drawn back at the corners?

(7) When a man sneers or snarls at another, is the corner of the upper lip over the canine or eye tooth raised on the side facing the man whom he addresses?

(8) Can a dogged or obstinate expression be recognised, which is chiefly shown by the mouth being firmly closed, a lowering brow and a slight frown?

(9) Is contempt expressed by a slight protrusion of the lips and by turning up the nose, and with a slight expiration?

(10) Is disgust shown by the lower lip being turned down, the upper lip slightly raised, with a sudden expiration, something like incipient vomiting, or like something spit out of the mouth?

(11) Is extreme fear expressed in the same general manner as with Europeans?

(12) Is laughter ever carried to such an extreme as to bring tears into the eyes?

(13) When a man wishes to show that he cannot prevent something being done, or cannot himself do something, does he shrug his shoulders, turn inwards his elbows, extend outwards his hands and open the palms; with the eyebrows raised?

(14) Do the children when sulky, pout or greatly protrude the lips?

(15) Can guilty, or sly, or jealous expressions be recognised? though I know not how these can be defined.

(16) Is the head nodded vertically in affirmation, and shaken laterally in negation?

Observations on natives who have had little communication with Europeans would be of course the most valuable, though those made on any natives would be of much interest to me. General remarks on expression are of comparatively little value; and memory is so deceptive that I earnestly beg it may not be trusted. *A definite description of the countenance under any emotion or frame of mind, with a statement of the circumstances under which it occurred, would possess much value* [my italics] .[26]

Darwin noted he received thirty-six answers to these questions which emphasise the meaning which can be inferred from descriptions of the physical movements of facial features in respect of certain emotional states. The answers he received formed the raw material of *The Expression of the Emotions*, even though he identified the emotion prior to asking for specifically denoted signs of the emotional expression under analysis.

What, then, can we make of Darwin's notes, questions, and observations? They were certainly anecdotal, collected in a rather haphazard manner with inconsistent results, but the question is whether they tell us anything about the development of Darwin's ideas (or indeed his position) on emotion and expression. One possible clue is in the last sentence of the instructions accompanying his questionnaire – 'a definite description of the countenance under any emotion or frame of mind, with a statement of the circumstances under which it occurred, would possess much value'. Fundamental to such an enterprise was an explanation of how the capacity to express emotion has evolved and why, as a result, expressions are associated with specific emotional states. To do this, Darwin needed to identify specific emotional states as well as their corresponding expressions and then map the development of common (or universal) expressions through groups of related organisms. If this could be done, many human emotions (like love, anger, and grief) and a number of allied instincts in man and the higher animals (such as laughter and perhaps even blushing) signal the existence of common evolutionary origins. If Darwin can convince us of this then we have to accept his argument about the continuity of the species. Having surveyed many of the writings on expression, and especially physiognomy – including James Parsons, Charles Le Brun, and Pieter Camper, as well as Lavater and Bell – Darwin concluded that the only possible explanation for the existence of some expressions in men as well as animals was that 'the structure and habits of all animals have been gradually evolved'. He went on:

The community of certain expressions in distinct though allied species as in the movements of the same facial muscles during laughter by man and by various monkeys, is rendered somewhat more intelligible, if we believe in their descent from a common progenitor. He who admits on general grounds that the structure and habits of all animals have been gradually evolved, will look at the whole subject of Expression in a new and interesting light. (p. 12)

The challenge for Darwin was to demonstrate how the instincts which express emotions might have progressed.

III

There were several writers in addition to Bell, Darwin wrote in the early pages of *The Expression of the Emotions*, whose works 'deserve the fullest consideration' – for instance, Moreau, Burgess, Duchenne, Gratiolet, and Piderit, together with Bain and Spencer – but he claimed the weakness of almost all these writers was their belief in the immutability of the species. 'All the authors who have written on Expression', he insisted, 'with the exception of Mr Spencer – the great expounder of the principle of Evolution – appear to have been firmly convinced that species, man of course included, came into existence in their present condition' (p. 10). Though his response to such resolute doctrinal belief was clear, he realised that the study of expression posed some difficult questions about human judgement and reason, not least because of the transience of emotional expressions and our over-familiarity with their physical movements. To investigate the properties of emotional states via facial expression was an almost impossible task, at once intractable and fascinating, and as he readily appreciated it brought with it particular problems:

The study of expression is difficult, owing to the movements being often extremely slight, and of a fleeting nature. A difference may be clearly perceived, and yet it may be impossible, at least I have found it so, to state in what the difference consists. When we witness any deep emotion, our sympathy is so strongly excited, that close observation is forgotten or rendered almost impossible; of which fact I have had many curious proofs. Our imagination is another and still more serious source of error; for if from the nature of the circumstances we expect to see any expression, we readily imagine its presence. (p. 13)

What better means of countering the distorting effects of sympathy or imagination than to present emotion as not transcendentally but physiologically purposeful? The advantage to Darwin of such an idea was that it offered an explanation of the possible mechanisms for the development of the mental, moral and social capacities of man and animals and, just as importantly, it gave emotion a structural function (in relation to reflex) with a profoundly physiological basis. This meant that the expression of the emotions, and the integrated mental and physical processes it involved, could be explained in terms of instinct, so acknowledging the functional relevance of animal traits to humans in direct opposition to the argument from design.

Throughout his investigations, Darwin looked for a description which would make clear the exact muscles involved in facial expression, the adaptive importance of the acts, and the resemblances between man and other animals (plates 17 and 18). To ensure the best possible foundations for this, he pinpointed six areas of interest which would, he hoped, help build up an accurate picture of the correspondence between the movements of the facial features (as well as physical gestures) and specific states of mind; these were the observation of infants, of the insane, of the facial muscles, of works of art, of the different races of men, and of animals. Whilst he found little of service in works of art, Darwin followed the lead of Bell in

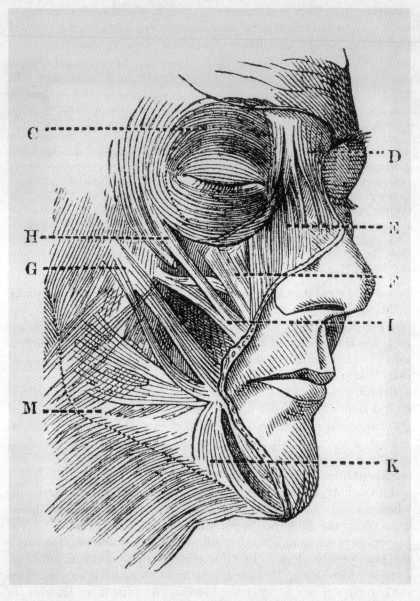

Plate 17   Charles Darwin, Diagram of facial muscles, from Henle, 1872.

Plate 18 Charles Darwin, Chimpanzee looking tired and sulky, by Mr. Wood, 1872.

taking particular notice of the many emotions expressed by infants; furthermore he was greatly assisted by Dr Duchenne, who galvanised the facial muscles of an old man and having photographed the expressions presented them to more than twenty people for their description of the emotion being expressed. Darwin was, in fact, heavily dependent on Duchenne for evidence of the actions of the facial muscles, as many of the plates in *The Expression of the Emotions* (e.g. plates 18 and 19) demonstrate, and he reproduced many of the photographs from Duchenne's *Mécanisme de la Physionomie Humaine* (1862). Duchenne derived his technique of galvanising facial muscles from the famous research of Luigi Galvani on animal electricity. Working in the last twenty or so years of the eighteenth century, Galvani proposed that the brain was a mechanism which generated and transmitted a nervous power that contained similar properties to electricity. Just as for Hartley the brain was made up in part of a puzzling substance called aether (a term borrowed, incidentally, from Newton), so for Galvani the brain produced a nerve fluid which could be compared to electricity.[27] The importance of the galvanisation of muscles of the face, for Darwin, was that it gave the understanding of expression an experimental dimension that it was previously closest to achieving with Bell. Darwin believed, in fact, that Duchenne made significant advances in localising muscles of the face (by stimulating the nerve endings with electric currents) and, more significantly, identifying those muscles which are involuntary, or as Darwin put it the 'muscles . . . least under the separate control of the will' (p. 5).

Three principles were proposed by Darwin to account for involuntary expressions of emotion in man and the lower animals – serviceable associated habits, antithesis, and the direct action of the nervous system. Although ostensibly they emerge out of his researches and observations along with the questionnaire (quoted above) which he distributed to friends and missionaries throughout the world, it will become clear that Darwin was heavily indebted to a number of leading thinkers from the scientific community of his day – William Carpenter, Thomas Laycock, Alexander Bain, and Herbert Spencer, to name a few – for the legitimation of his ideas on the instincts for expressing emotions. The starting-point for Darwin's first principle was, however, Bell's omission of an explanation for why expressions are associated with specific emotions. Do we learn such associations or is this knowledge inherited (through instinct)? During the course of his early observations on expression in Notebooks M and N, Darwin held to habit as the mechanism of evolutionary change. His explanation was indebted to his grandfather Erasmus Darwin, as William Montgomery has shown, and established around the idea that the practice of intentional actions and behaviour would, over a long period of time, result in those actions becoming inherited and therefore instinctive.[28] This was Darwin's problem: if emotions are states of mind producing expressions in response to certain events, what was the origin of the emotional expression? That is to say, how did the connections between emotions and expressions evolve? Driven by a conviction of the uselessness of expressions of emotion, Darwin proposed that the complex actions of expression were sometimes serviceable as a means of relieving specific states of mind, and that the repetition of the state of mind, through habitual association, causes the related expression to be produced. 'Some actions ordinarily associated through habit with certain states of the mind may be partially repressed through the will', he wrote, 'and in such cases the muscles which are least under the separate control of the will are the most liable to act, causing movements which we recognise as expressive' (p. 28). For example, the man who rubs his eyes in confusion or coughs when embarrassed will probably continue to perform these actions, 'as if he felt a slightly uncomfortable sensation in his eyes or windpipe', whenever he experiences those emotions (p. 32). Moreover, this kind of connection or association between expression and emotion could be inherited within the man's family and passed through successive generations.[29]

To be sure, Darwin's understanding of the sequence dictating this associative process was as follows: emotion is the mental response to an external stimulus and expression is its physical action. Where the emotion produces an expressive action the physical state of the individual is affected and if the emotion is repeatedly experienced this association can become a habitual one. The existence of 'serviceable associated habits', which Darwin's first principle recognises, was intended to show how some intentional acts refer both to specific events and emotional states and through repeated and habitual use might be associated continually with that emotional state. However, his second and third principles were concerned with those instinctive expressions that did not originally emerge through habit. The second principle was antithesis and, as with the first principle, Darwin claimed there was a connection between certain states of mind and particular habitual actions. In this case, though, he proposed an inverse relation between states of mind and habitual actions, so that when a certain emotion was induced there was a tendency to perform movements of an opposite or contrary nature. The most common example of this in humans is shrugging, where the physical action of moving the shoulders upwards is supposed to represent helplessness and, rather ironically, an inability to act. But given the difficulty of distinguishing conventional from innate expressions in humans, Darwin says, his attention will be focussed on the lower animals. Take the following example:

We will now turn to the cat. When this animal is threatened by a dog, it arches its back in a surprising manner, erects its hair, opens its mouth and spits . . . The attitude is almost exactly the same as that of a tiger disturbed and growling over its food, which every one must have beheld in menageries. The animal assumes a crouching position, with the body extended; and the whole tail, or the tip alone, is lashed or curled from side to side. The hair is not in the least erect. Thus far, the attitude and movements are nearly the same as when the animal is prepared to spring on its prey, and when, no doubt, it feels savage. But when preparing to fight, there is this difference, that the ears are closely pressed backwards; the mouth is partially opened, showing the teeth; the forefeet are occasionally stuck out with protruded claws; and the animal occasionally utters a fierce growl.

Let us now look at a cat in a directly opposite frame of mind, whilst feeling affectionate and caressing her master; and mark how opposite is her attitude in almost every respect. She now stands upright with her back slightly arched, which makes the hair appear rather rough, but it does not bristle; her tail, instead of being extended and lashed from side to side, is held quite stiff and perpendicularly upwards; her ears are erect and pointed; her mouth is closed; and she rubs against her master with a purr instead of a growl. (pp. 56–7)

The point is that certain movements are used despite their uselessness – the actions of the cat are linked with a specific state of mind and the occurrence of a contrary state seems to produce actions of an opposite kind. In the above example, then, the cat crouches in position when ready to spring in anger whereas it adopts an upright and arched posture when it feels affectionate. The correlation between anger and affection was reasoned through analogy, on the basis of which Darwin claimed that emotions (or sensations) of an opposite kind produced expressive acts of a similarly opposite kind.

Darwin's third principle was concerned with the action of the nervous system on the instincts expressing emotion, independently of the will and habit. The excitement of the sensorium evokes nervous energy which 'is transmitted in certain directions, dependent on the connection of the nerve-cells, and, as far as the muscular system is concerned, on the nature of the movements which have been habitually practised' (p. 66). He had first suggested something like this in the M notebook when he said that the excess of certain emotions caused a physical change in the nerve cells that was transmitted through the whole body. But Darwin is indebted to Spencer for outlining the implications of this idea, and quotes

Plate 19   Charles Darwin, Seven photographs of man and child, looking unhappy;
from Oscar Rejlander, 1872.

directly from the second series of his *Essays, Political and Speculative*
(1863): 'overflow of nerve-force, undirected by any motive, will
manifestly take the most habitual routes; and, if these do not suffice,
will next overflow into the less habitual ones'.[30] Habitual actions of
this sort would, Darwin insisted, alter the pathways of the nerves in
the same way that the continual use of a footpath, for instance,
wears grooves into its route. In a similar vein, as sustained repetition
of certain ideas – such as names, dates, or events – eventually
commits them to memory, so habit stands here for something akin to
emotional (or what is sometimes called muscular) memory. Many
examples could be cited from *The Expression of the Emotions* from the
trembling of muscles and the release of perspiration out of fear and
anxiety to expressions of pain and excitement, but it is the description of the stages of grief which stands out:

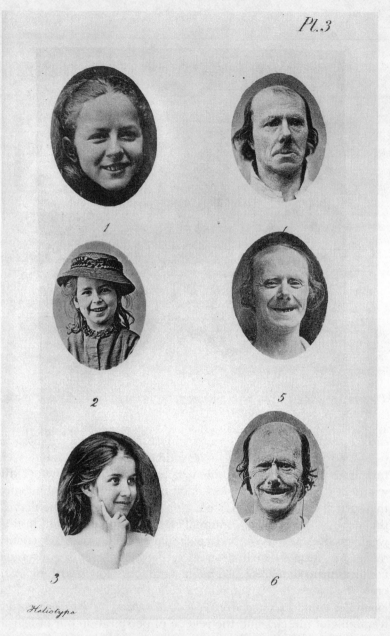

Plate 20   Charles Darwin, Six photographs of man and child, looking happy; from
G. B. Duchenne and Oscar Rejlander, 1872.

When a mother suddenly loses her child, sometimes she is frantic with grief, and must be considered to be in an excited state; she walks wildly about, tears her hair or clothes, and wrings her hands [1]. This latter action is perhaps due to the principle of antithesis, betraying an inward sense of helplessness and that nothing can be done. The other wild and violent movements may in part be explained by the relief experienced through muscular exertion, and in part by the undirected overflow of nerve-force from the excited sensorium [2]. But under the sudden loss of a beloved person, one of the first and commonest thoughts which occurs, is that something might have been done to save the lost one . . . As soon as the sufferer is fully conscious that nothing can be done, despair or deep sorrow takes the place of frantic grief. The sufferer sits motionless, or gently rocks to and fro; the circulation becomes languid; respiration is almost forgotten, and deep sighs are drawn [3]. All this reacts on the brain, and prostration soon follows with collapsed muscles and dulled eyes [4]. An associated habit no longer prompts the sufferer to action, he is urged by his friends to voluntary exertion, and not to give way to silent, motionless, grief. Exertion stimulates the heart, and this reacts on the brain, and aids the mind to bear its heavy load. (pp. 84–5; my numbers)

This passage emphasises the instincts which guide the associated expressions of grief, and visual illustration is provided in the photographs taken by Duchenne and Rejlander which Darwin chose to represent this emotional state (plate 20).[31] The description conveys the consequences of nervous action in terms which clarify the causal relationship between the idea of grief [1] with its related emotional state [2] and the physical action [3] and expression [4]. We learn what expressions are used for each stage of the emotional experience, how they are produced, and when they occur. There seem, in fact, to be two stages of grief: the first is the mental awareness of the emotion [1] and [2], and the second is the associated physical response [3] and [4]. The physiological explanation of this principle is vague, and Bain pointed this out to Darwin in a review of *The Expression of the Emotions*,[32] but the underlying assumption is the uselessness of an excess of nervous energy except for its tendency towards expression.

IV

What linked all of Darwin's principles was the axiom that habits might become instincts. Take, for example, the involuntary shutting of the eyelids even when the surface of the eye is not touched. This movement is only indirectly useful as a means of expressing a certain emotional state, usually a fear of being hurt, but it does express directly the influence of habitual association in linking an expressive action with an emotion, as we will tend to respond like this whenever fear and pain are mixed together. This is Darwin on the subject:

Another familiar instance of a reflex action is the involuntary closing of the eyelids when the surface of the eyes is touched. A similar winking movement is caused when a blow is directed towards the face; but this is an habitual and not a strictly reflex action, as the stimulus is conveyed through the mind and not by the excitement of the peripheral nerve. The whole body and head are generally at the same time drawn suddenly backwards.[33]

The involuntary action of the eyelids was due, he maintained, to habit rather than reflex in that the recall of this particular sensation in the eye is enough to cause the shutting of the eyelids. Darwin's aim was to show how the habitual actions of an individual can produce structural changes in the organisation not just of the individual but also the species. 'Every one understands', he had pronounced in *The Origin of Species*, 'what is meant when it is said that instinct impels the cuckoo to migrate and to lay her eggs in other birds' nests':

An action, which we ourselves should require experience to enable us to perform, when performed by an animal, more especially by a very young one, without any experience, and when performed by many individuals in the same way, without their knowing for what purpose it is performed, is usually said to be instinctive. (p. 169)

What he had not recognised fully until working on *The Expression of the Emotions* was the extent to which habit might be responsible for structural modifications as, for instance, with the overflow of nervous energy which emerged in inherited form as instinct.[34]

The possibility that physical structures could transmit instincts through generations had transpired in 'Notebook M' where he reflected on the consequences of age, and certain kinds of mental impairment, for the operation of thought.[35] As Richards has compellingly shown, the correlation of thought with brain structure 'would indeed suffice for Darwin's purpose, which in this instance was twofold':

First, the supposed causal relationships of brain to mind could furnish a perfectly naturalistic (and evolutionary) explanation for apparently non-biological mental traits. But second, these relationships could also be marshalled in more instrumental ways, as mechanisms of evolutionary change.[36]

It is here that the concept of memory looms large, enabling Darwin to draw an analogy between involuntary muscular movement and unconscious memory: 'when a muscle is moved the motion becomes habitual & involuntary', he mused in his 'Notebook M', 'when a thought is thought very often it become habitual & involuntary, that is involuntary memory . . . an intentional recollection of anything is solely by association, & association is probably a physical effect of the brain'.[37] The interpretation of thought and memory as natural functions of the brain allows Darwin to conceive of memory passing into inherited instinct and in so doing to posit the biological transmission of altered brain structures via reproduction, in the same way that other physical traits are passed through generations. 'These facts showing what a train of thought, action, etc. will arise from physical action on the brain', he said, 'render much less wonderful the instincts of animals'.[38] Richards argues, in fact, that what he calls the 'habit-instinct mechanism', as well as environ-mental effects, should be seen alongside sexual production as equally important adaptive mechanisms for species change in Darwin's evolutionary system.[39]

Darwin's view of expressions as instincts was derivative of a number of different sources. By making memory the analogue for instinct Darwin was advocating a mechanism of inheritance prox-imate to the Lamarckian notion of the inheritance of acquired characteristics. It is well known that much of Darwin's thought on the heritability of instinct was determined by Lamarck's theory.[40] Lamarck drew on popular beliefs such as the idea that a blacksmith whose forearms developed as a result of his labours would produce children, grandchildren, great-grandchildren and so on, with an inbuilt tendency to develop the same muscles. Lamarck systematised this sort of folklore, maintaining that the repetition of particular actions would result in the instinct for these actions being passed through generations. One of the best summaries of this Lamarckian theory of the mechanism of instinct is given by Stephen Jay Gould in *Ontogeny and Phylogeny* (1977):

The general form of the argument was simple and acceptable to all adherents: the acquisition of character is like learning; since characters so acquired are inherited in proportion to the intensity of their producing stimuli, inheritance is like memory (learning is retained through memory; memory is enhanced by constant repetition over long periods; actions invoked at first by conscious thought become automatic when repeated often enough). Instincts are the unconscious remembrance of things learned so strongly, impressed so indelibly into memory, that the germ cells themselves are affected and pass the trait to future generations.[41]

There is, however, another body of knowledge in the physiological and psychological writings of the time which Darwin drew upon. One of the main references throughout his work on instinct was Johannes Müller's *Handbuch der Physiologie des Menschen für Vorlesungen* (1834–40) and it is clear from the marginal annotations to his copy of an English translation that he gleaned many thoughts on inherited instincts from Müller.[42] There is also an important connection between Darwin and the theories of action developed by his English contemporaries, William Carpenter and Thomas Laycock. Both Carpenter and Laycock were interested in the functional basis of physical action, with the former particularly interested in the development of the physiological basis for judgement and reason, and the latter in the integration of mind and body through the reflex concept.[43] It is clear that Darwin had read and was familiar with the work of these men, and judging from the references in *The Expression of the Emotions*, their theories of reflex action (together with those of Bain and Spencer) seem to have legitimated Darwin's explanation of the instinct expressing emotion.[44]

Carpenter extended the notion of reflex action to include an explanation of its increasingly complex function in the organic world. In an article 'On the Voluntary and Instinctive Actions in Living Beings', he advanced a vision of the world wherein reflex action was found in its lowest form in the vegetable kingdom and in its highest in the sensory-motor response of certain creatures in the animal kingdom.[45] Throughout, the principle of the continuity of nature was maintained on the basis that even a vegetable was capable of a primitive form of adaptation which was the foundation of subsequent and higher progression. An established and well-respected figure in the scientific community of the mid-century, due largely to his publication of many books (and textbooks) in comparative physiology and the physiology of the mind, Carpenter sought to prove that physiology could become as law-like as the physical sciences. There were, he thought, two assumptions which could be made about vital phenomena: in the first place they were the product of properties of organised tissues which acted according to regular laws; and in the second these properties were intrinsic to phenomena but only developed as a result of certain circumstances: '[the properties] were not superadded to matter in the process of organisation; . . . but this act calls out or developes [*sic*] the properties which previously existed in the particles subjected to it, but which are not manifested except under the peculiar circumstances which this new disposition of them produces'.[46] His argument in respect of instinct was that amongst the majority of animals the actions which were called 'instinctive' ought to be 'comprehended in the same general definition' and also extended to include those functions which animals held in common with vegetables.[47] The adaptations of animals to stimuli produced different internal mechanisms, thus allowing for the development of a nervous system with a greater or lesser degree of complexity. 'The successively more complex nervous actions of animals', Clarke and Jacyna contend, 'were to be depicted as development from a primitive functional type, just as the structures that mediated them were complicated versions of the simple ganglion. Throughout he maintained that the "reflex function is the simplest application of the Nervous System in the animal body".'[48] The two main classes of action initially proposed by Carpenter – the most basic being adaptation of an organism to its environment and the more sophisticated being the response of an animal to stimuli through action but without sensation – were later amplified to include what he called 'consensual acts', that is 'those which are the direct result of sensations acting immediately on the motor nerves without the intervention of volition'.[49]

Carpenter is an intriguing example because, in parallel with Hartley and Bell, he allowed for an explanation of the laws governing the organic world without removing the Creator (or higher being) from the picture. Though he was not an associationist (as he believed in the agency of the will in mental process), Carpenter attempted to hold a developmental conception of nervous function and action in balance with dualist assumptions about the existence of the soul and the agency of the will. Nonetheless, by the publication of the fourth edition of his *Principles of Human Physiology* (1853) some of his ideas had been significantly revised, not least in response to Laycock's thoroughly monist understanding of brain function. A lecturer in clinical medicine at York Medical School until 1855, and subsequently professor of medicine at Edinburgh,

Laycock focussed his researches on the relation between neural and emotional illness. Central to this relation was the concept of reflex action, which, he argued, operated directly through the nervous system. In opposition to the idea that the cerebral hemispheres had different structures and functions, he proposed that reflex action was responsible for the functional homogeneity both within an individual organism and between groups of related organisms. Denying the validity of Cartesian dualism, he argued for a fundamental continuity between the nervous organisation of humans and animals, and emphasised the importance of an integrated view of psychical and vital phenomena. Crucial to this was the relationship he posited between human and animal action:

The law of the unity of type and function in animals, applied . . . to the function of the cerebro-spinal axis in man, has shown . . . that the transmission of structure and function is gradual, and consequently no strong line of demarcation can be drawn between the manifestations of its various functions. The automatic act passes insensibly into the reflex, the reflex into the instinctive, the instinctive are *quasi* emotional, the emotional intellectual.[50]

An action, he contends, is the product of a complex and involuntary causal reaction which starts with an automatic response and then moves in rapid stages from the reflex to the instinctive and the emotional and finally the intellectual. 'If the cerebral ganglia be but a higher development of the spinal', he explained,

the medullary and the cortical substance must correspond to the white and grey matter of the cord, and if it be acknowledged, (as indeed has been proved beyond question) that a combined action of sets of muscles, exhibiting a design of conservation may be developed in the spinal cord without the aid of volition, how can we deny the same qualities to the encephalic ganglia, or in other words, to the cerebral hemispheres and their connexions?[51]

There are obvious parallels in this passage with the associationist psychology of Hartley and the physiological writings of Bell, especially in the prominence given to the nervous system in determining physical as well as mental action although, unlike them Laycock suggested a progressive explanation of the nervous structures of different types of animals.

This is the setting from which Darwin sought an explanation of how instincts expressing emotion evolved. He believed that some of our expressions originated in movements which, although no longer useful to us, were once of use to our common ancestors. This is illustrated by the epigraph from Spencer which opens the present chapter. Specific patterns of behaviour, like bees making cells or cuckoos migrating to a foreign nest, became instinctive and so set the foundations for evolutionary change. One of the most interesting examples Darwin gives of behavioural acts controlled by habit, where the frequent repetition resulted in significant changes in the brain that could become inherited, was an experiment on his own young children:

I shook a pasteboard box close before the eyes of one of my infants, when 114 days old, and it did not in the least wink; but when I put a few comfits into the box, holding it in the same position as before, and rattled them, the child blinked its eyes violently every time, and started a little. It was obviously impossible that a carefully-guarded infant could have learnt by experience that a rattling near its eyes indicated danger to them. But such experiences will have been slowly gained at a later age during a long series of generations; and from what we know of inheritance, there is nothing improbable in the transmission of a habit to the offspring at an earlier age than that at which it was first acquired by the parents.[52]

In sum, experience is unimportant in explaining the involuntary responses of an infant to stimuli when compared to the building of habit through successive generations. Having shown the relationship between emotions and expressions, and related why we express our feelings through bodily actions, Darwin affirmed the habitual and instinctive character of emotional expressions by drawing, in particular, on contemporary physiological ideas of action. Using the three principles of habit, antithesis, and nerves, he investigated the purpose of expressions and their associated mental and physical states but there remained the question of the biological function of expressions of emotion. Could the evolutionary origins of the instincts expressing emotion prove the communicative value (and hence universality) of expressions? The answer can be found, I suggest, in Darwin's explanation of blushing.

v

Of all expressions, blushing was, to Darwin, the most peculiar to humans; it would, he said, 'require an overwhelming amount of evidence to make us believe that any animal could blush'.[53] It was also a fundamental and imponderable subject, and Darwin set out to prove that blushing was a physical and mental act in the same

way as grief or laughter. As such, it served as a further illustration
of instinctive action as a mechanism for evolutionary change, albeit
in a social form. It was the social aspect of blushing which most
intrigued Darwin, for it provided the opportunity to clarify and
expand the theory of moral evolution he had advanced in *The
Descent of Man*. The crux of the matter was the moral sense (or
conscience) because he deemed it the most important element of
human nature, that which was responsible for the constitution of
society through the sympathetic identification between individuals.
The essence of this argument was advanced in the first part of *The
Descent of Man*, in which Darwin suggested that there was evidence
for the common development of man and the higher apes in
primitive mammalian stock. The homologous structures of men
and apes, especially vertebral structures, and their susceptibility to
the same diseases, such as 'hydrophobia, variola, the glanders, &c',
'proves the close similarity of their tissues and blood, both in
minute structure and composition' (p. 11). There is even, Darwin
continues, a marked similarity in the courting, mating, and nur-
turing rituals of men and apes: 'monkeys are born in almost as
helpless a condition as our own infants; and in certain genera the
young differ fully as much in appearance from the adults, as do our
children from their full-grown parents' (p. 13). Of most interest to
Darwin, though, were the rudimentary organs of man: the ability
to twitch the forehead or ears, the hairy shoulders of some men,
the smaller molars, veriform appendix, and human tailbone. Only
the evolutionary explanation of origins makes these traits explicable, for it suggests that they had a more perfect form in man's
ancestors, and that although they had a clearly developed function
for early men, a change in habits rendered them largely useless yet
still heritable in some forms. In an analysis of the mental and
moral character of man, Darwin tries 'to see how far the study of
the lower animals can throw light on one of the highest psychical
faculties of man' (p. 71).

Using his experiences on the *Beagle*, in particular, Darwin maintained that the difference in mental power between men and the
lower animals, though in some cases considerable, is one of degree
not kind; and as with the mind so with morals:

nor is the difference slight in moral disposition between a barbarian, such
as the man described by the old navigator Byron, who dashed his child on
the rocks for dropping a basket of sea-urchins, and a Howard or Clarkson;

and in intellect, between a savage who does not use any abstract terms, and a Newton or Shakespeare. (p. 35)

There are many reasons why this should be so – and the faculties of attention, memory, imagination, and reason are the examples Darwin cites – but the moral sense or conscience reveals by far the most significant difference between man and the lower animals. The evolution of the moral sense is complex, Darwin argued, but the following is the most plausible summary: 'that any animal whatever, endowed with well-marked social instincts, would inevitably acquire a moral sense or conscience, as soon as its intellectual powers had become well-developed, or nearly as well developed, as in man' (pp. 71–2). There were, he said, four interrelated stages in the evolution of moral sense, starting with the development of social instincts in our ancestors and ending with the building of habit. The first stage is the emergence of social instincts in animals which causes them to group together in mutual pleasure and sympathy for the benefit of each other. Such social groups or communities would depend, Darwin said, on the interrelation or close association of individuals within a species (but would not extend to all individuals within a species). The second stage is dependent upon the progress of the mental powers and occurs when members of this newly founded society have sufficient intellect to discriminate past actions and experiences in terms of their attendant feelings; so, for instance, 'that feeling of dissatisfaction which invariably results . . . from any unsatisfied instinct, would arise, as often as it was perceived that the enduring and always present social instinct had yielded to some other instinct, at the time stronger, but neither enduring in its nature, nor leaving behind it a very vivid impression' (p. 72). The third stage was distinguished by the acquisition of language which would enable each individual within the community to express their needs and desires, bearing in mind, of course, the good of the collective mass. Hence, the 'guide to action' would be determined by the common opinion of the public good. The final stage sees habit become increasingly important in directing the conduct and behaviour of individuals, especially in acting for the common good.

The sequence of these four stages is markedly instinctual in character, for Darwin believed that the mistake of previous work on the moral sense – and in particular that of James Mackintosh and Alexander Bain – was in seeing social and sympathetic reactions as learned associations, whether the response was to close relatives or complete strangers. How, he asks, can sympathy be 'excited in an immeasurably stronger degree by a beloved than by an indifferent person'? The answer is simple: where sympathy may once have been associative response, it is now quite clearly an instinct, 'for we are led by the hope of receiving good in return to perform acts of sympathetic kindness to others; and there can be no doubt that the feeling of sympathy is much strengthened by habit' (p. 82). The moral sense is invoked by Darwin to indicate those feelings which provide the individual with a capacity for sympathetic identification with another, and also an ability to choose between certain opposing instincts: it prompts an individual to perform certain actions, regulating the pleasure or pain which accompanies the actions, and yet it is, above all, a social instinct which emerges out of the relations between animals.[54] Emphasising the actions of the social group over that of the individual, therefore, Darwin looked to the longevity of the species to explain action rather than to states of pleasure and pain, and so viewed altruistic acts as those intended for the general good and, therefore, unlikely to increase the happiness of individuals. 'As he had in his early theory of morals', Richards explains, 'Darwin based his argument for the instinctive nature of human sympathy largely on analogy: social and altruistic responses in animals were instincts, so why should they not also be in that most social of animals?'[55]

Though the social instinct was crucial to the formation of the moral sense, the highest psychical faculty of man, it was not sufficient on its own to develop the possibility of right conduct and action; intelligence was also required. Darwin said, as quoted above, that for any animal in possession of 'well-marked social instincts' the emergence of a moral sense was directly related to the development of its intellectual powers. So whereas social instincts would stimulate an individual to action, an evolved intellect was required if reason and experience as well as the power to discriminate were to contribute to the higher conduct of a moral being. 'Why should a man feel he ought to obey one instinctive desire rather than another', Darwin asks, 'why does he bitterly regret if he has yielded to the strong sense of self-preservation, and has not risked his life to save that of a fellow-creature; or why does he regret having stolen food from severe hunger?'[56] The reason is that the more enduring social instincts prevail over the less persistent ones, and a moral being (a human being and not an animal) is an individual who can compare past and future actions with approval or disapproval. It is worth recalling Hartley's use of the moral sense here; the connection of instinct, reason, and morality was immensely significant for him because it enabled actions, affections, dispositions, and judgements to be strung together as the vital constituents of human nature, the determinants of what it is to be human. What Darwin did was to reconfigure the moral sense so that it stood for a biological impulse which had its origins in the social instinct and so maintained the continuity principle whilst also indicating the gradations of scale between the higher and lower animals. It is, moreover, in terms of moral action and obligation that a progressive distinction between highness and lowness was made – a long quotation from *The Descent of Man* serves to illustrate the point:

If it be maintained that certain powers, such as self-consciousness, abstraction, &c., are peculiar to man, it may well be that these are the incidental results of other highly-advanced intellectual faculties; and these again are mainly the result of the continued use of a highly developed language. At what age does the new-born infant possess the power of abstraction, or become self-conscious and reflect on its own existence? . . . The moral sense perhaps affords the best and highest distinction between man and the lower animals . . . the social instincts – the prime principle of man's moral constitution – with the aid of active intellectual powers and the effects of habit, naturally lead to the golden rule, 'As ye would that men should do to you, do ye to them likewise'; and this lies at the foundation of morality. (p. 106)

One of the most manifest signs of the distinctive qualities of man was blushing, which Darwin used to advance a theory of self-consciousness.

In the early pages of Darwin's 'Notebook N', for instance, an entry on the act of blushing is enclosed between two statements on the methodology of science. It is worth quoting the passage in full because it introduces the notion of science as a metaphysical activity, and also indicates the relevance of blushing to the universality of expressions. Shifting from assertion to speculation, and back to assertion, Darwin mused:

All science is reason acting <<systematizing>> on principles, which even animals practically know <<art precedes science – art is experience & observation. – >> in balancing a body & an ass knows one side of triangle shorter than two. V. Whewell. Induct. Sciences – Vol I p.334
. . .

Does a negress blush. – I am almost sure Fuegia Basket did. & Jemmy, when Chico plagued him – Animals I should think would not have any emotion like blush. – when extreme sensation of heat shows blood is pumped over whole body. – is it connected with surprise. – heart beginning to beat – children inherit it <<ins>> like instinct, preeminently so – who can analyse the sensation, when meeting a stranger. who one may like. dislike, or be indifferent about, yet feel shy. – not if quite stranger. – or less so. –

When learning facts for induction. one is obliged carefully to separate its memory from all ordinary lines of association. – is totally distinct from learning it by heart. Do not our necessary notions follow as consequences on habitual or instinctive assent to propositions, which are the result of our senses, or our experience. – Two sides of a triangle shorter than third. is this necessary notion, ass has it. – (author's emphasis)[57]

These passages are problematic because they work at cross-purposes, making manifest a desire to work out both a scientific methodology and the physiological mechanism of blushing. It is quite clear that by defining science as reason, system, and principles, Darwin is grappling for an understanding of what kind of observations are necessary to constitute a scientific explanation and what kind of distinctions are sufficient to qualify this. Science, he states, succeeds art because art is intuitive, emotional, and experiential whereas science is logical, factual, and abstract. The challenge is to educate mind and memory to organise facts into a system based on metaphysical principles which is capable of judgement and objective understanding. Some context is, however, necessary for his remarks on man.

The Fuegians were, Darwin thought, capable of great civility; an observation he based on his acquaintance with the three Fuegians – Jemmy Button, York Minster, and Fuegia Basket – who were due to be returned to their homeland by the captain of the *Beagle*, Robert FitzRoy, accompanied by a church-funded missionary, Richard Matthews. Tierra del Fuego, literally the land of fires, was a wild and desolate place which in Darwin's eyes quickly became inseparable from the savage, even primitive people he encountered there. The *Beagle* crew arrived in Tierra del Fuego on 17 December 1832. It was just about the most southerly point of South America, and a strong impression of it remained embedded in Darwin's mind for many years ahead as he was fascinated by the state and behaviour of the Fuegians – they were physically menacing, tall, dark, semi-naked, with faces painted with red and black stripes, and they were also skilled mimics. 'I could not', he wrote, 'have believed how wide was the difference, between savage and civilized man. It was greater than between a wild and domesticated animal, in as much as in man there is a greater power of improvement.'[58] The hostility was tangible, but what interested Darwin in particular was the correspondence between the indigenous people and their 'civilized' compatriots. Could they really be fellow men? Surely the civilizing process had a beneficial effect on Jemmy, York, and Fuegia, had it not? Darwin was considerably perplexed. Here was a land and a type of people which did not fit easily into his developing view of nature; 'Death, instead of Life, seemed the predominant spirit'.[59] What was the place of these people in nature? Man may now appear to be an incredibly varied species, but there was no evidence to suggest that the Fuegians were more like animals than he was. In other words, primitive races like the Fuegians were not inherently different from Europeans and, as Jemmy, York, and Fuegia had proved through their capacity for improvement, the difference was of degree rather than kind. The enquiry 'does a negress blush' thus affirms the direction of Darwin's thoughts on the descent of man together with the accessibility of emotional expression to scientific analysis. A subsequent notebook entry cuts to the heart of the issue:

Blushing is intimately connected with thinking of ones appearance, – does the thought drive blood to surface exposed, face of man, face, neck– <<upper>> bosom in woman: like erection. shyness is certainly very much connected with thinking of oneself. – <<blushing>> is connected with sexual, because each sex thinks more of what another thinks of him, than any one of his own sex. – Hence, animals. not being such thinking people. do not blush. – sensitive people apt to blush. – The power of vivid mental affection, on separate organs most curiously shown in the sudden cures of tooth ache before being drawn.[60]

The act of blushing is motivated by a self-consciousness – 'thinking of ones appearance . . . thinking of oneself' – which stimulates the physiological mechanism of the body and arouses the sexual instinct. There are also shades of Spencer in the alignment of appearance with character; Darwin makes explicit the correspondence between personal appearance and mental character in so far as being conscious of one's physical self is, in fact, an awareness of one's inner mental self. Moreover, the emphasis on personal and physical appearance leads him to suggest that blushing is intimately bound up with the relations of the sexes.

The penultimate chapter of *The Expression of the Emotions* is devoted to blushing under the heading, 'Self-attention – Shame – Shyness – Modesty: Blushing'. The point of the chapter and its importance to an evolutionary account of expression is spelt out in the following terms:

The belief that blushing was *specially* designed by the Creator is opposed to the general theory of evolution, which is now so largely accepted . . . Those who believe in design, will find it difficult to account for shyness being the most frequent and efficient of all causes of blushing, as it makes the blusher to suffer and the beholder uncomfortable, without being of the least service to either of them. They will also find it difficult to account for negroes and other dark-coloured races blushing, in whom a change of colour in the skin is scarcely or not at all visible. (p. 335)

The point is that the argument from design, which posits a world wherein everything has purpose, does not make clear what place a seemingly useless instinct to express emotion, such as a blush, has. Yet, to Thomas Burgess in *The Physiology or Mechanism of Blushing* (1839), blushing was one of the 'involuntary acts of the mind upon the vital organs and their several functions' which emphasised in particular 'the physical changes that are produced in different parts of the body'.[61] Burgess, a physiologist and physician whose treatise on blushing was written around the same time as Darwin was recording his thoughts and observations in the notebooks 'M' and 'N', introduced his subject in personal and rhetorical terms:

Who has not observed the beautiful and interesting phenomenon of BLUSHING? Who indeed has not had it exemplified in his own person, either from timidity during the modest and sensitive days of boyhood, or from the conscious feeling of having erred in maturer years? When we see the cheek of an individual suffused with a blush in society, immediately our sympathy is excited towards him; we feel as if we were ourselves concerned, and yet we know not why. The condition by which the emotion thus proclaimed is excited, viz., extreme sensibility, the innate modesty and timidity which are the general concomitants of youth, enlist our feelings in favour of the party, appeal to our better nature, and secure that sympathy, which we ourselves may have claimed from others on similar occasions.[62]

A number of things are worth noting in this extract. The first is the observation that blushing is an experience which is shared by all human beings; the second is that it appears to be a social emotion (or instinct) which is roused in the company of others; and the third is its appeal to our feelings of identification in sympathising with the afflicted. Under what conditions, then, do we blush? Furthermore, what parts of our body blush and how does it happen?

By concentrating on effects rather than causes, Burgess claimed that blushing involved both physical and moral conditions; that is to say, an internal sense of excitement and an external sense of approbation. It was the embarrassment which a blush often makes poignantly visible that seems to have fascinated Burgess most, for it signalled a causal relationship between the sensorium and the heart, the blood-vessels, and the nerves. Hence, the purpose of blushing was twofold: to illustrate shame and so check the moral faculties, and also to demonstrate the physiological mechanisms of the body. In the opinion of Burgess, the first of these always took precedence over and controlled the second, and so even though blushing took many forms and had many different varieties – including so-called true blushes attendant upon moral causes, conscience and feeling, and false blushes produced by a morbid sensibility, shame, and rage – what marked it out for investigation was the essence of its action:

Is it not probable that it was with this intention [as an illustration of shame] the Creator of man endowed him with this peculiar faculty of exhibiting his internal emotion, or more properly speaking, of the internal emotions exhibiting themselves, for no individual blushes voluntarily; it would, therefore, appear to serve as a check on the conscience, and prevent the moral faculties from being infringed upon, or deviating from their allotted path.[63]

The sense of this statement is as follows: if blushing is a higher emotional expression, exclusively human and specifically moral, then it must be intended to illustrate the fundamental discontinuity between man and other animals, and to prove that human beings are a separate, divinely created species. This argument is similar to those of Lavater and Bell. Indeed, Burgess was adamant that the possession of the moral sense was the highest mark of man (and man alone) and was, therefore, absolutely crucial to the capacity to blush. That animals cannot blush is implied in the above quotations; however, Burgess admits both that 'savage man' possesses the faculty of reason and that it is likely that he possesses the capacity to blush, albeit in a primitive social form.

Like Burgess, Darwin suggests that the moral sense is an index of mental development, and like Spencer he implies that there is a connection between physical appearance and the relations of the sexes. Unlike Burgess, however, he does not see blushing as evidence of the moral sense, nor as an innate quality, and unlike Spencer he does not claim that variation arises from imperfection. Blushing, once a direct and now an indirect result of attention, may not always be visible or self-evident in all races and certainly is not evident in animals, but it is, nonetheless, the product of habit:

The moral nature of man has reached the highest standard as yet attained, partly through the advancement of the reasoning powers and consequently of a just public opinion, but especially through the sympathies being rendered more tender and widely diffused through the effect of habit, example, instruction, and reflection. It is not improbable that virtuous tendencies may through long practice be inherited. With the more civilised races, the conviction of the existence of an all-seeing Deity has had a potent influence on the advancement of morality. Ultimately man no longer accepts the praise or blame of his fellows as his chief guide, though few escape this influence, but his habitual convictions controlled by reason afford him the safest rule. His conscience then becomes his supreme judge and monitor. Nevertheless the first foundation or origin of the moral sense lies in the social instincts, including sympathy; and these instincts no doubt were primarily gained, as in the case of the lower animals, through natural selection.[64]

Consequently, blushing was, to Darwin, the product of consciousness above conscience: it was not evidence of the design of a higher being but proof of the importance of an individual's attention to emotional expression; moreover, the purpose of this experience is primarily physiological. 'The reddening of the face from a blush is due to the relaxation of the muscular coats of the small arteries, by which the capillaries become filled with blood; and this depends on the vaso-motor centre being affected', he disclosed in *The Expression of the Emotions.*

No doubt if there be at the same time much mental agitation, the general circulation will be affected; but it is not due to the action of the heart that the network of minute vessels covering the face becomes under a sense of shame gorged with blood. We can cause laughing by tickling the skin, weeping or frowning by a blow, trembling from the fear of pain, and so forth; but we cannot cause a blush, as Dr. Burgess remarks, by any physical means, – that is by any action on the body. It is the mind which must be affected. Blushing is not only involuntary; but the wish to restrain it, by leading to self-attention actually increases the tendency. (p. 310)

This is an interesting passage as it demonstrates the fallacy of physical control over the body: blushing is an involuntary response to certain mental stimuli and any attempt to curb it increases rather than lessens the effect.

Many examples are given by Darwin of blushing, illustrating its propensity to occur in the young and in women, but what united those individuals who blush, across sexual and racial lines, was an acute sense of being observed, primarily in their appearance but also in their (moral) conduct. As a result, there were certain social situations wherein blushing occurs – shame, shyness, and modesty – which were, Darwin claimed, demonstrative of a confused mind. He explained:

From the intimate sympathy which exists between the capillary circulation of the surface of the head and of the brain, whenever there is intense blushing, there will be some, and often great, confusion of mind. This is frequently accompanied by awkward movements, and sometimes by the involuntary twitching of certain muscles. (p. 344)

This localisation of attention causes a localised response which explains why, when we blush, the sensation is felt to be greatest on the face and head. Above all, it is a regard for the opinions of others which explains why an individual blushes and how their offspring can acquire this instinct. Through generations, the repeated action of the sensory nerves which cause blushing become habitual actions which respond to any kind of attention (real or imaginary) and cause this distinctive response whether others are thinking of us or not:

The hypothesis which appears to me the most probable, though it may at first seem rash, is that attention closely directed to any part of the body tends to interfere with the ordinary and tonic contraction of the small arteries of that part. These vessels, in consequence, become at such times more or less relaxed, and are instantly filled with arterial blood . . . Whenever we believe that others are depreciating or even considering our personal appearance, our attention is vividly directed to the outer and visible parts of our bodies; and of all such parts we are most sensitive about our faces, as no doubt has been the case during many past generations. Therefore, assuming for the moment that the capillary vessels can be acted on by close attention, those of the face will have become eminently susceptible. Through the force of association, the same effects will tend to follow wherever we think that others are considering or censuring our actions or character. (p. 336)

There is nothing metaphysical here; simply a reassertion of the evolutionary argument and an example of the first principle of expression. With blushing, then, we are presented with a rather distinctive emotional experience: the process by which the sensory and motor nerve cells are connected occurs as a direct result of attention on a specific part of the body, usually the face.

VI

'All the chief expressions exhibited by man are the same throughout the world.' With this statement, Darwin moved towards the conclusion of his study of expression. 'This fact is interesting', he continued, 'as it affords a new argument in favour of the several races being descended from a single parent-stock, which must have been almost completely human in structure, and to a large extent in mind, before the period at which the races diverged from each other' (p. 355). The context for this fact should be familiar to us by now; that is to say, *The Expression of the Emotions* was intended to prove the sameness (or universality) of the main expressions like fear, love, and anger both in man and animals. But, as I have shown, the ramifications of Lavater's teachings on the instinctive nature of physiognomic judgements become clear in the hands of Darwin. The importance of this connection ought not to be overlooked, for descriptions of human nature of the kind given by Lavater were inextricably linked with complex arguments about the actions of the nerves and the muscles, the efficacy of the will, and the purpose of sensations and emotions. It is, of course, the last of these – the purpose of emotions – which has most concerned us in this chapter, because whilst Lavater (and Bell) maintained the transcendental function of instincts expressing emotion within a theological world view, Darwin insisted on their biological function without fully recognising their communicative value.

That said, *The Expression of the Emotions* reconfigures our understanding of expression, providing a means of studying the originating sequences which determine emotion, habit, instinct and behaviour. The examination of the physiological basis of grief and blushing, in particular, align Darwin with the tradition of physiology and psychology which comes from Hartley and moves in different terms through the work of Bell, Carpenter, and Laycock to Bain and Spencer. An evolutionary account of expression seems to provide precisely the natural scientific explanation of man and mind which Lavater could not offer: that is to say, though refuting the transcendental grounds for the physiognomic study of expression, Darwin's enquiry is actually underwritten by physiognomic ideas about the innate (and essential) quality of emotional expression. Grief, for example, was seen by Darwin as innate and instinctive, and so as much a product of evolutionary change (through inheritance) as of physiological structure. The hardest thing for Darwin was how to explain the evolution of the instinct expressing emotion; only, he proclaimed, by accepting descent as the 'hidden bond of connexion which naturalists have been seeking under the term of the natural system' can emotion be presented in naturalistic terms.[65]

CHAPTER 6

The promise of a new psychology

In the distant future I see open fields for far more important
researches. Psychology will be based on a new foundation, that
of the necessary acquirement of each mental power and
capacity by gradation. Light will then be thrown on the origin
of man and his history.

Charles Darwin[1]

An understanding of character was crucial to descriptions of human
nature throughout the nineteenth century in so far as it required
discrimination of the essential properties of an individual or what
makes them who they are. As we have already seen, a description of
human nature for Darwin necessarily involved consideration of the
moral sense, which was, he insisted, a strain of social instinct. The
feeling which prompted altruistic acts, and also caused pain when
obligation was ignored, was responsible for the formation of groups
of animals into social wholes, Darwin contended, enabling mutual
cooperation and functional activity within the grouping and allowing
it to evolve through the operation of natural selection at a higher
level than that of the individual. The thesis depended, though, on
the selective process; the change in social groups (or species) as a
result of natural selection occurred through the evolution of instinctive responses intended to preserve the welfare of the community,
and so those best able to express their moral impulses and perform
behaviour appropriate for the benefit of the group survived.

This was the basic model from which Francis Galton developed
his science of social improvement, drawing also on Lamarckian ideas
about inheritance.[2] His eugenics, meaning the consideration of
which human beings were 'good in stock, hereditarily endowed with
noble qualities',[3] were intended to improve the race of man. The
object was, he said, 'to take note of the various hereditary faculties of
different men, and of the great differences in families and races';

to learn how far history may have shown the practicability of supplanting human stock by better strains, and to consider whether it might not be our duty to do so by such efforts as may be reasonable, thus exerting ourselves to further the ends of evolution more rapidly and with less distress than if events were left to their own course.[4]

It was well known that farmers and gardeners could create permanent breeds of animals and plants which were strong in particular characteristics, and Galton thought human stock could be improved in the same way; thus the strength of particular character was achieved through breeding.

The premise of Lamarck's theory was that the inheritance of acquired characteristics was an adaptive mechanism which distorted the trend towards a linear and parallel progression of animal and plant forms.[5] But whereas plants were directly affected by a new environment, growing different structures adapted to the new place which were then passed on through generations, animals were indirectly affected by a change in their physical conditions, and adaptation only occurred at an individual level with some new habits being formed and some organs being used more than others. Lamarck maintained that these individual responses might be transmitted through generations until the point at which they became inbuilt and produced a permanent change in the species; that is to say, the influence of the environment caused hereditary material to be reformulated. This Lamarckian theory directed Darwin's early speculations on evolution which were based around the inheritance of acquired characteristics, and though he later became convinced of the randomness of variation from a growing familiarity with the principles of breeding, he left open the possibility of the inheritance of acquired characteristics having a supporting role (to natural selection) in the evolutionary process.

Placed between the evolutionary thought of his cousin Darwin and Gregor Mendel's science of genetics, Galton's work represents an important juncture in the use of natural selection to explain intellectual progress.[6] His aim was to construct a system of social selection based on breeding – in effect the foundation of the eugenics movement – which would extend Darwin's mechanisms of natural selection and habit to their logical conclusion.[7] Central to this was his study of heredity – started in the 1860s and continuing throughout his life – which investigated the transmission of mental ability in human beings through a statistical method of analysis. To

Darwin, the moral sense was the index of the evolution of the mind as it most clearly demonstrated the instinctive capacity to act for the good of a group over the individual. Yet Darwin considered mental ability to be more or less equivalent within specific groups, like the one formed by his contemporaries at Cambridge, and only hard work and sustained effort might cause slight intellectual variations or differences. The merit of Galton is to have recognised quite large variations in the intelligence of what he considered to be civilized men, which meant that an explanation of the progress of their minds could be brought within the remit of the evolutionary process and its adaptive mechanisms. His first published essay, 'Hereditary Talent and Character' (1865), was intended to show the heritability of intellect: a pronounced mental talent would, he said, be evident within families and through generations, because an aptitude for science, mathematics, law, literature, and painting was the result of the biological transmission of mind and character. The evidence provided was interesting because Galton examined a number of sources, including the biographical dictionaries of distinguished people, the roll of past presidents in the British Association, and the lists both of lord chancellors and of senior classics at Cambridge, and found that the men noted often had quite close relatives who were also of pronounced intellect. The implication was clear: 'when a parent has achieved great eminence, his son will be placed in a more favourable position for advancement, than if he had been the son of an ordinary person'.[8]

A sense of the fundamental importance of the intellect was first advanced by Herbert Spencer in *Principles of Psychology* (1855) in the form of the law of mental progress (or intelligence):

It is a dominant characteristic of Intelligence, viewed in its successive stages of evolution, that its processes, which, as originally performed, were not accompanied with a consciousness of the manner in which they were performed, or of their adaptation to the ends achieved, become eventually both conscious and systematic . . . Thus children reason, but do not know it. Youths know empirically what reason is, and when they are reasoning. Cultivated adults reason intentionally, with a view to certain results. The more advanced of such presently inquire after what manner they reason. And finally, a few reach a state in which they consciously conform their reasonings to those logical principles which analysis discloses.[9]

From the simplest organism in its environment to the distinction of individuals into more complex species via intelligence, Spencer believed that the development of mental life involved a gradual movement from reflex action to instinct, memory, reason, and will: 'the evolution of life is an advance in the Speciality of the correspondence between internal and external relations'.[10] His evolutionary synthesis depended on a number of assumptions: the similarity of life and mind; the equivalence of mental actions, distinguished only by the differing complexity of internal and external relations; and the need for organisms to adapt to an environment which changed frequently through the modification of their internal organisation according to external factors:

> Every form of Intelligence being, in essence, an adjustment of inner to outer relations; it results that as, in the advance of this adjustment, the outer relations increase in number, in complexity, in heterogeneity, by degrees that cannot be marked: there can be no valid demarcations between the successive phases of Intelligence.[11]

Mind was, for Spencer, the product of organic development rather than a unique gift bestowed by a creator, and though life and mind evolved alongside each other, the difference between mental and physical responses meant that the nervous system and eventually a distinctive form of life, namely intelligence, were produced.[12]

Both Darwin and Galton shared this conviction, as we have seen – their views on the heritability of intellect were distinguishable though not incompatible – yet contemporary ideas about the development of intellect and the moral sense (or conscience) were diverse (and often contentious) as scientists, philosophers, and theologians dealt with the ramifications of evolutionary thought.[13] In *The Geological Evidences of the Antiquity of Man* (1863), for example, Charles Lyell suggested that talent was arbitrarily bestowed on individuals, occurring as the result of a sudden leap in the progress of mind: 'the birth of an individual of transcendent genius, of parents who have never displayed any intellectual capacity above the average standard' was therefore comparable to the shift from 'unprogressive intelligence of the inferior animals' to the 'improvable reason manifested by Man'.[14] Galton opposed this idea and its transcendental explanation of genius, arguing instead for the inheritance of talent through generations in the same way as biological traits were transmitted. 'What an extraordinary effect might be produced on our race if its object was to unite in marriage those who possessed the finest and most suitable nature, mental, moral, and physical!', he exclaimed.[15] The only problem was the behaviour of individuals, since a 'craving for drink, or for gambling, strong sexual passion, a proclivity to pauperism, to crimes of violence, and to crimes of fraud' had the effect of transmitting the vices of men along with their virtues. There was, then, a natural curb on the heritability of intellect with a negative influence on the selection of genius, but Galton believed this could be countered by the artificial selection of those individuals with the finest mental, moral, and physical nature, thus manufacturing higher social groups. The fact was the existence of relatively few men of good character was 'more adverse to early marriages than is natural bad temper, or inferiority of intellect'.[16] Galton's theory of the improvement of the human race could be achieved, he supposed, by controlling its breeding, a notion he derived from Spencer and, of course, Malthus.[17]

The origins of some of these ideas can be found in the arguments about healthy stocks discussed in chapter four, and in particular the notion that a superior physical beauty was the expression of higher mental development – a subject broached in an essay on 'Personal Beauty' (1854) by Spencer himself. For Galton, the most important elements of these discussions of the improvement of healthy stocks were the possibilities they suggested for a change in social attitudes and policy. 'The moral and intellectual wealth of a nation largely consists in the multifarious variety of the gifts of the men who compose it, and it would be the very reverse of improvement to make all its members assimilate to a common type.'[18] Galton believed that the vital clue to supporting the wealth of a nation, and so maintaining the healthiest of stocks, was each and every individual's face. In an important section on facial features in *Inquiries into Human Faculty* (1883), he explained the significance of the variety of expression:

The difference in human features must be reckoned great, inasmuch as they enable us to distinguish a single face among those of thousands of strangers, though they are mostly too minute for measurement. At the same time, they are exceedingly numerous. The general expression of a face is a multitude of small details, which are viewed in such rapid succession that we seem to perceive them all at a single glance. If any one of them disagrees with the recollected traits of a known face, the eye is quick at observing it, and it dwells upon the difference. (p. 4)

Facial difference indicates individuality but it points also to the fit between an individual expression and a common type. What Galton sought was a means of distinguishing individual from type so that he could map the heritability of intellect through generations and demonstrate the transmission of mental, moral, and physical nature as something akin to biological traits. He acknowledged the essential weakness of existing modes of analysis (and discrimination) in a crucial passage:

The physiognomical difference between different men being so numerous and small, it is impossible to measure and compare them each to each, and to discover by ordinary statistical methods the true physiognomy of a race. The usual way is to select individuals who are judged to be representatives of the prevalent type, and to photograph them; but this method is not trustworthy, because the judgement itself is fallacious. It is swayed by exceptional and grotesque features more than by ordinary ones, and the portraits supposed to be typical are likely to be caricatures. (pp. 5–6)

The problem is twofold: firstly, how can the physiognomic development of the human race be measured; and secondly, on what grounds can physiognomical differences between races be identified? Galton's problem, like Lavater's, was the adequacy of the standards against which difference might be delineated, but whereas Lavater reduced particular observations of expressions to common types, Galton wanted to make difference explicable in terms of the statistical mean of a specific sample.

To Galton, the coherence of a race depended on the existence of an ideal type to which individuals might bear some resemblance and from which they might deviate. In a system like Galton's (and indeed Spencer's) where the imperative was improvement towards the ideal – the healthiest stocks, the most suitable races, the best strains of blood – the aim was to select out as many examples of inferior variations as possible so as to allow the superior forms to flourish. 'The face and the qualities it connotes probably give a clue', he postulated, 'to the direction in which the stock of the English race might be improved'. And he continued:

It is the essential notion of a race that there should be some ideal typical form from which the individuals may deviate in all directions, but about which they chiefly cluster, and towards which their descendants will continue to cluster. The easiest direction in which a race can be improved is towards that central type, because nothing new has to be sought out. It is only necessary to encourage as far as practicable the breed of those who conform most nearly to the central type, and to restrain as far as may be the breed of those who deviate widely from it. Now there can hardly be a more appropriate method of discovering the central physiognomical type of any race or group than that of composite portraiture. (pp. 14–15)

The composite method appeared to offer a way out of the idiosyn-crasies of observation which impacted upon the study of expression (and were manifest in Lavater's teachings). That is to say, the use of composites removes the study of expression from instinctive grounds and proposes a rather more scientific approach based on mathema-tical process: it formulates a method of observation which mediates between individual and type, superimposing a series of photographs of facial expressions upon one another and using a succession of fractional exposures to produce a composite image. Galton's compo-site method involved a number of stages and worked like this: first, collect the full-face photographic portraits; second, reduce the portraits to the same size; third and fourth, secure the portraits and superimpose them upon each other; fifth, place a sensitive camera inside the book; and sixth, photograph each page of the book without moving either the camera or the plate: 'so that', in Galton's words, 'an image of each of the portraits in succession was thrown on the same part of the sensitised plate' (p. 9). If this sounds strange, the results were even stranger.

The collection of images given by Galton are divided into eight different composite types – Alexander the Great; two sisters; six members of one family; twenty-three Royal Engineers; fifteen cases of tubercular disease; twelve cases of criminal types; fifty-six con-sumptive cases; and one hundred and fifty non-consumptive cases – under the categories of personal and family, health, disease, crimi-nality, consumption and other maladies, respectively. Amidst ears which appear to float in odd places and some very blurred, almost indistinguishable features, the composite portrait which was pro-duced was meant to illustrate the relation of individual expressions to a common type and in this way illustrate the degree (not kind) of physiognomic difference between individuals. What is remarkable about the composites produced by this method is the extent to which Galton claimed they form 'an ideal composition' because they make manifest a specific inherited trait:

The effect of composite portraiture is to bring into evidence all the traits in which there is agreement, and to leave but a ghost of a trace of individual peculiarities. There are so many traits in common in all faces that the composite picture when made from many components is far from being a blur; it has altogether the look of an ideal composition. (p. 10)

As he said elsewhere, composites were 'pictorial statistics' intended to give us 'generic pictures of man':

They are the pictorial equivalents of those elaborate statistical tables out of which averages are deduced. There cannot be a more perfect example than they afford, of what the metaphysicians mean by generalisations, when the objects generalised are objects of vision, and when they belong to the same typical group, one important characteristic of which is that medium characteristics should be far more frequent than divergent ones.[19]

At issue in using composites to picture descent as the transmission of particular characteristics was its verification of the study of expression; according to Alan Sekula, 'in effect Galton believed that he had translated the Gaussian error curve into pictorial form'.[20]

Throughout *Inquiries into Human Faculty*, Galton delves into the physical differences of race, offering psychometric experiments with which to record the operations of mind, and suggesting the causes that hinder the 'unlimited improvement of highly-bred animals' (p. 306). The delicacy of constitution and the infertility of 'highly-bred animals' contributed, he suggested, to limiting the improvement of human stock, but melioration can be found in the variation of intelligence. Thus, man can support the natural order of things 'by furthering the course of evolution. He may use his intelligence to discover and expedite the changes that are necessary to adapt circumstance to race and race to circumstance, and his kindly sympathy will urge him to effect them mercifully' (pp. 334–5). To be sure, Galton constructed a system of judgement which allowed the practical application of physiognomic teachings to assume a primary role in ordering and regulating society, by emphasising human intelligence. His ideas on resemblance, hereditary transmission, and unity of type affirmed the place of physiognomy in describing human nature and character whilst also helping to reconfigure its future. For as suggested throughout this book, the emergence of scientific rationales for the expression of the emotions placed the study of expression on new ground. The epigraph to this chapter indicates where that ground might be for Darwin, and points the study of 'mental power and capacity', particularly in man, towards psychology.

The appeal of physiognomy was that it made many of the topics of psychological study, like mental ability, emotional experience, and behaviour, not only comprehensible but also accessible, via a system designed to instruct an individual on how to translate their largely instinctive responses to actual appearances into a practical knowledge of the relationship between expression and emotion. In sum, it explained human nature in terms of a uniform order of types or kinds and it did so at the level of the individual, translating particular observations into general theories of the nature of emotion and character. As the previous chapter has shown, however, Darwin's evolutionary thought transformed what was involved in expression from a description of character into an explanation of how the instinct to express emotions has evolved within a specific group or community. Even so, though the materialism of Darwinian thought confounded physiognomy, the biological explanation of emotional expression had a distinct Lamarckian dimension in its understanding of the building of habit through inheritance. This contradiction proved important for Galton, as his writings are exemplary of both these positions: whilst he used physiognomic teachings to direct his thought, he also radically rewrote many of its assumptions about mind, character, and behaviour. The science of social improvement he proposed mediated an understanding of the essential qualities of character with a belief in the capacity, even necessity, for artificially inducing progress, especially the progress of the intellect. The science of mind promised by Lavater here reaches a definitive turning-point, for Galton saw the survival of the species as dependent on the selection of its healthiest and most agreeable types; that is to say, those beneficial to the betterment of man. 'The new mental attitude is one of a greater sense of moral freedom, responsibility, and opportunity', he wrote; 'the new duty which is supposed to be exercised concurrently with, and not in opposition to the old ones upon which the social fabric depends, is an endeavour to further evolution, especially that of the human race'.[21]

Notes

INTRODUCTION

1 Johann Caspar Lavater, *Essays on Physiognomy; for the Promotion of the Knowledge and the Love of Mankind*, trans. Thomas Holcroft, 3 vols. (London, 1789–93), I, pp. 16–17.

2 Ibid., p. 17.

3 *The Encyclopaedia Britannica*, 4th edn, 20 vols. (Edinburgh: Archibald Constable & Co., 1810), XVI, p. 439.

4 The period of time I have identified here is, of course, a rough one which is mapped out between the first and last publications which have a direct bearing on physiognomy. Though Lavater's work was first published in German in 1775–8, *Physiognomische Fragmente, zur Beförderung der Menschenkenntniß und Menschenliebe*, 4 vols. (Leipzig and Winterthur: Weidmann, Reich, and Steiner, 1775–8), I am taking Holcroft's edition of 1789–93 as the starting-point for this study. See chapter one, n. 11 for further details on the editions. Galton's work is an important one for physiognomy because it marks a turn from an understanding of expression as representative of character towards an understanding of expression as part of an emerging psychological tradition. Though Galton's work is not the first to point out such uses for physiognomy, it is, I think, highly significant for its application of an extreme kind of evolutionary thought – namely, the selection of the healthiest stocks – to physiognomic teachings.

5 The Human Genome Project is the most prominent example of recent attempts to examine the genetic constituents of human nature and so provide a map of the different forms of chromosomes which determine the structure of our DNA.

6 John Stuart Mill, 'Bain's Psychology' (1859), *Essays on Philosophy and the Classics*. I use the term 'psychology' quite deliberately here to denote a branch of scientific knowledge which has as its subject the notion of subjectivity itself. Psychology has a problematic position within the field of science precisely because it attempts to theorise that which is familiar to us all as human beings; namely, our common understanding of what

it is to be an individual with mind and body. This can be traced, as Roger Smith has argued, to the fact that the history of psychology has been difficult to reconstruct (and therefore often neglected), as historians of science have looked mainly at the experimental psychology in Germany and the United States of the 1870s onwards as the application of a method to the themes of psychology. See *Inhibition: History and Meaning in the Sciences of Mind and Brain* (California: University of California Press, 1992). My interest here is in the broadest of parallels between physiognomy and psychology as sciences of man and mind.

7 One of the earliest philosophical treatises on physiognomy is a work thought to be written by Aristotle, entitled *Physiognomics*, which represents probably the first attempt to present physiognomy as an interpretative process with a specific and extensive methodology. For a comprehensive account of the classical tradition of physiognomy, see Elizabeth C. Evans, 'Physiognomics in the Ancient World', *Transactions of the American Philosophical Society* 59 (1969): 5–97.

8 See John Graham, 'The Development of the Use of Physiognomy in the Novel', PhD diss., Johns Hopkins University, 1960; 'Lavater's Physiognomy: A Checklist', *Papers of the Bibliographical Society of America* 15 (1961): 308; 'Lavater's Physiognomy in England', *Journal of the History of Ideas* 22 (1961): 561–72; 'Character Description and Meaning in the Romantic Novel', *Studies in Romanticism* 5 (1966): 208–18; *Lavater's 'Essays on Physiognomy': A Study in the History of Ideas* (Bern: Peter Lang, 1979).

9 Graeme Tytler, *Physiognomy in the European Novel: Faces and Fortunes* (Princeton: Princeton University Press, 1982), p. xv.

10 See for instance, Maria Allentuck, 'Fuseli and Lavater: Physiognomical Theory and the Enlightenment', *Studies in Voltaire and the Eighteenth Century* 15 (1967), 89–112; Christopher Rivers, *Face Value: Physiognomical Thought and the Legible Body in Marivaux, Lavater, Balzac, Gautier, and Zola* (Madison: University of Wisconsin Press, 1994); Ellis Shookman, ed., *The Faces of Physiognomy: Interdisciplinary Approaches to Johann Caspar Lavater* (Columbia: Camden House, 1993); Michael Shortland, 'Barthes, Lavater and the Legible Body', *Economy and Society* 14 (1985): 273–312; Judith Weschler, *A Human Comedy: Physiognomy and Caricature in Nineteenth-Century Paris* (Chicago: University of Chicago Press, 1982). Some of these works will be analysed in more detail in chapter one.

11 Susan F. Cannon, *Science in Culture: The Early Victorian Period* (New York: Dawson and Science History Publications, 1978); Ludmilla Jordanova and Roy Porter, *Images of Earth: Essays in the History of the Environmental Sciences* (Chalfront St. Giles: The British Society for the History of Science, 1979); David Knight, *The Age of Science: The Scientific World-View in the Nineteenth Century* (Oxford: Basil Blackwell, 1996); James Moore, ed., *History, Humanity and Evolution: Essays for John C. Greene* (Cambridge: Cambridge University Press, 1989).

12 See Adrian Desmond, *The Politics of Evolution: Morphology, Medicine, and*

Reform in Radical London (Chicago: University of Chicago Press, 1989), p. 7.

13 There is a huge literature on these debates, but the works with the greatest direct relevance to this study include: Gillian Beer, *Darwin's Plots: Evolutionary Narrative in Darwin, George Eliot, and Nineteenth-Century Fiction* (London: Ark Books, 1983); F. W. Burkhardt, *The Spirit of the System: Lamarck and Evolutionary Biology* (Cambridge, Mass.: Harvard University Press, 1977); John W. Burrow, *Evolution and Society: A Study in Victorian Social Theory* (Cambridge: Cambridge University Press, 1966); William Coleman, *Biology in the Nineteenth Century: Problems of Form, Function, and Transformation* (New York: John Wiley, 1972); Charles C. Gillispie, *Genesis and Geology: A Study in the Relations of Scientific Thought, Natural Theology, and Social Opinion in Great Britain, 1790–1850* (Cambridge, Mass.: Harvard University Press, 1951); Robert M. Young, *Darwin's Metaphor: Nature's Place in Victorian Culture* (Cambridge: Cambridge University Press, 1985).

14 Mary Cowling, *The Artist as Anthropologist: The Representation of Type and Character in Victorian Art* (Cambridge: Cambridge University Press, 1989), p. xix.

15 See chapter two for an account of the connections between natural theology and the emergence of physiological ideas. See also John C. Greene, *Science, Ideology, and World View: Essays in the History of Evolutionary Thought* (Berkeley: University of California Press, 1981) and Frank Miller Turner, *Between Science and Religion: The Reaction to Scientific Naturalism in Late Victorian England* (New Haven: Yale University Press, 1974).

16 Thomas Kuhn, *The Structure of Scientific Revolutions*, 2nd edn (Chicago, 1970). On questions relating to the demarcation of science from non-science, Imre Lakatos and Alan Musgrave's edited collection, *Criticism and the Growth of Knowledge* (Cambridge: Cambridge University Press, 1970), is extremely useful.

17 For a more detailed explanation of Kuhn's thesis about the rationality of theories of scientific method, see W. H. Newton-Smith, *The Rationality of Science* (London: Routledge & Kegan Paul, 1981).

18 One of the leading proponents of a contextualist approach to science is Robert M. Young who, since his early work in the 1970s, has consistently argued for the importance of an historiographical discourse in writing on nineteenth-century science, in particular. He proposed that 'ideas do not beget ideas but that people do so in particular historical contexts and that the meaning of those ideas is exquisitely bound to the particularity of those contexts'. See *Darwin's Metaphor: Nature's Place in Victorian Culture* (Cambridge: Cambridge University Press, 1985), p. 176. Young's view is not without its problems as it often neglects to discuss the extraordinary range of ideas in the favoured 'common context', but it has, nonetheless, been influential on the subsequent generations of historians of science. In the last twenty or so years there have been a

number of collections which reflect on the meanings of science and its practical applications in the Victorian period: see, for example, Barry Barnes and Steven Shapin, eds., *Natural Order: Historical Studies of Scientific Culture* (London: Sage, 1979); James Paradis and Thomas Postlewait, eds., *Victorian Science and Victorian Values: Literary Perspectives* (New York: New York Academy of Sciences, 1981); Ludmilla Jordanova, ed., *Languages of Nature: Critical Essays on Science and Literature* (London: Free Association Books, 1986); Patrick Brantlinger, ed., *Energy and Entropy: Science and Culture in Victorian Britain* (Bloomington: Indiana University Press, 1989). Most recently, Bernard Lightman has suggested that 'by ordering nature to conform to a particular pattern, scientists and intellectuals frequently reveal the social order for which they yearn; and in the process of practicing science, of measuring, experimenting, and controlling phenomena, we not only find nature but also encounter ourselves as inquisitive, social, and political beings'. Bernard Lightman, ed., *Victorian Science in Context* (Chicago: University of Chicago Press, 1997).

19 Alison Winter, 'The Construction of Orthodoxies and Heterodoxies in the Early Victorian Life Sciences', *Victorian Science in Context*, ed. Lightman, pp. 24–50, p. 25.

20 See chapter one for a detailed discussion of the various criticisms levelled at Lavater's work.

21 There is not a huge literature on phrenology – the definitive account is Roger Cooter's, *The Cultural Meaning of Popular Science: Phrenology and the Organisation of Consent in Nineteenth-Century Britain* (Cambridge: Cambridge University Press, 1985), but see also D. de Giustino, *Conquest of the Mind: Phrenology and Victorian Social Thought* (London: Croom Helm, 1975). There has been more interest in mesmerism and the spiritualist tradition in recent years – see especially Alison Winter, *Mesmerised: Powers of Mind in Victorian Britain* (Chicago: Chicago University Press, 1998).

22 Winter, 'The Construction of Orthodoxies', p. 43.

23 William Whewell, *Philosophy of the Inductive Sciences* (1840), 2nd edn, 2 vols. (London: J. W. Parker & Sons, 1847), p. 8. The companion volume to this was the *History of the Inductive Sciences*, 3 vols. (London, 1837). The former volume was reprinted twice in the twenty years following its first appearance. The third edition (1858–60) appeared as three separate but related works: *History of Scientific Ideas*, *Novum Organon Renovatum*, and *The Philosophy of Discovery*. A selection from these works is available in a modern edition, first published in 1968 and recently reprinted: *William Whewell, Theory of Scientific Method*, ed. Robert E. Butts (Indianapolis: Hackett Publishing Co., 1989).

24 Whewell, *Quarterly Review* 51 (1834): p. 59.

25 See Richard Yeo, 'William Whewell, Natural Theology and the Philosophy of Science in Mid-Nineteenth-Century Britain', *Annals of Science* 36 (1979): 493–516.

26 Whewell, *Astronomy and General Physics considered with reference to natural theology*, (1833), 5th edn (London: William Pickering, 1836), p. 304. For the context to the development of Whewell's scientific methodology, see Jonathan Smith, *Fact and Feeling: Baconian Science and the Nineteenth-Century Literary Imagination* (Madison: University of Wisconsin Press, 1994) and Richard Yeo, *Defining Science: William Whewell, Natural Knowledge, and Public Debate in Early Victorian Britain* (Cambridge: Cambridge University Press, 1993).

27 Whewell, *Philosophy of the Inductive Sciences*, p. 6.

28 Ibid., p. 12.

29 Ibid., p. 73.

30 Lewis Wolpert, *The Unnatural Nature of Science* (London: Faber and Faber, 1992), p. 17.

31 John Dupré, *The Disorder of Things: Metaphysical Foundations of the Disunity of Science* (Cambridge, Mass.: Harvard University Press, 1993), pp. 1–2 and 6–7.

32 Dupré, *The Disorder of Things*, p. 37.

33 See Roger Smith, 'The Background of Physiological Psychology in Natural Philosophy', *History of Science*, 11 (1973): 75–123. See also the dissertation from which this article is drawn: 'Physiological Psychology and the Philosophy of Nature in Mid-Nineteenth-Century Britain', PhD diss., University of Cambridge, 1970, and also his invaluable *Fontana History of the Human Sciences* (London: HarperCollins, 1997).

1   A SCIENCE OF MIND? THEORIES OF NATURE, THEORIES OF MAN

1 John Cross, *An Attempt to Establish Physiognomy upon Scientific Principles. Originally delivered in a series of lectures* (Glasgow: A & J. M. Duncan *et al.*, 1817), pp. 6–7.

2 James Parsons, 'Human Physiognomy Explain'd: In the Crounian Lectures on Muscular Motion', *Philosophical Transactions of the Royal Society* 44 (1747): 1–82.

3 For a broader historical explanation of this shift from the eighteenth into the nineteenth century, see Karl Figlio, 'Theories of Perception and the Physiology of Mind in the Late Eighteenth Century', *History of Science* 12 (1975): 177–212, and 'The Metaphor of Organization: An Historiographical Perspective on the Bio-Medical Sciences of the Early Nineteenth Century', *History of Science* 14 (1976): 17–53; Christopher Fox, 'Defining Eighteenth-Century Psychology: Some Problems and Perspectives', *Psychology and Literature in the Eighteenth Century*, ed. Christopher Fox (New York: AMS Press, 1987); Gary Hatfield, 'Remaking the Science of Mind', *Inventing Human Science*, ed. Christopher Fox, Roy Porter, and Robert Wokler (Berkeley: University of California Press, 1995); June Goodfield Toumlin, 'Some Aspects of

English Physiology, 1780–1840', *Journal for the History of Biology* 2 (1969): 283–320.

4 Roger Smith, *Inhibition: History and Meaning in the Sciences of Mind and Brain* (Berkeley: University of California Press, 1992), p. 35. On a similar theme, see also Smith's 'The Human Significance of Biology in the Nineteenth Century', *Nature and the Victorian Imagination*, ed. U. C. Knoepflmacher and G. B. Tennyson (Berkeley: University of California Press, 1977).

5 Parsons, 'Human Physiognomy Explain'd', p. 2.

6 Ibid., p. 32.

7 Ibid., pp. 33–4.

8 Ibid., p. 47.

9 See in particular L. S. Jacyna, 'Immanence or Transcendence: Theories of Life and Organization in Britain, 1790–1835', *Isis* 74 (1983): 311–29.

10 The first publication in English was a five-volume edition, edited by Henry Hunter, with the following title: *Essays on Physiognomy: Designed to Promote the Knowledge and Love of Mankind*, trans. Henry Hunter, 5 vols. (London, 1789). Hunter's edition was lavishly illustrated and beautifully presented, and as a result was priced at the princely sum of thirty guineas. Thomas Holcroft produced a cheaper (and shorter) edition which was reissued more than eighteen times over the next eighty years (even though several other editions were produced in the 1790s): *Essays on Physiognomy; for the Promotion of the Knowledge and the Love of Mankind; Written in the German Language by J. C. Lavater, and translated into English by Thomas Holcroft*, 3 vols. (London, 1789–93). As it is the more popular edition of Lavater's work, I shall refer to Holcroft's edition throughout this book. For a complete list of the publications of Lavater's *Essays on Physiognomy*, see John Graham, 'The Development and Use of Physiognomy in the Novel', PhD diss., Johns Hopkins University, 1960, appendix C. An abbreviated version of the publication history can be found in Graham's article, 'Lavater's *Physiognomy* in England', *Journal of the History of Ideas* 22 (1961): 561–72, p. 562.

11 Jacyna, 'Immanence or Transcendence', p. 312.

12 Jennifer Montagu, 'Le Brun's Lecture on Expression', *The Expressions of the Passions: The Origin and Influence of Charles Le Brun's 'Conférence sur l'expression générale et particulière'* (New Haven: Yale University Press, 1994), p. 126. My account of Le Brun is indebted to Montagu's excellent translation of the text of Le Brun's lecture on expression, and her detailed explanation of the institutional setting and reception of Le Brun's work.

13 See Montagu, *The Expressions of the Passions*, appendix I, pp. 141–3.

14 Roger de Piles, *Cours de Peinture par Principes* (Paris, 1708). I quote from the English translation: *The Principles of Painting, translated into English by a painter* (London, 1743). See also Thomas Puttfarken, *Roger de Piles' Theory of Art* (New Haven: Yale University Press, 1985).

15 Montagu, 'Le Brun's Lecture on Expression', p. 126.

16 See Christopher Rivers, *Face Value: Physiognomical Thought and the Legible Body in Marivaux, Lavater, Balzac, Gautier, and Zola* (Wisconsin: University of Wisconsin Press, 1994), for a reading of the influence of Descartes on Le Brun in the context of the history of physiognomy.

17 René Descartes, 'Discourse on Method', *The Philosophical Works of Descartes*, trans. John Cottingham, Robert Stoothoff, and Dugald Murdoch, 2 vols. (Cambridge: Cambridge University Press, 1985), I, p. 127.

18 The value and significance of Descartes' theory of mind and body have been well-documented. See in particular John Cottingham, *Descartes* (Bristol: Thoemmes Press, 1996); Bernard Williams, *Descartes, the Project of Pure Enquiry* (Harmondsworth: Penguin, 1978).

19 Descartes, 'The Passions of the Soul', *The Philosophical Works of Descartes*, I, pp. 328–9.

20 Norman Bryson discusses this aspect of Le Brun's studies of expression in *Word and Image: French Painting of the Ancien Regime* (Cambridge: Cambridge University Press, 1981).

21 Le Brun's theory was not confined to the interpretation of human form but made a series of comparisons between man and animals. Indeed, there is a striking similarity between Le Brun's analogues of men and animals and Giovanni Battista della Porta's study of man and animal physiognomy in his *Della fisionomia dell'huomo*, which was translated into French in 1655 and 1665 and is very likely to have been familiar to Le Brun. On this connection, see Montagu, *The Expressions of the Passions*, pp. 20–30.

22 Montagu, 'Le Brun's Lecture on Expression', p. 126.

23 *Ibid.*, p. 128.

24 Ernst Gombrich used this image of a dial to characterise Le Brun's theory of expression. It was, he claimed, a 'dial theory' because the body and especially the face should manifest the hidden actions or movements of the soul. See 'Moment and Movement in Art', *The Image and the Eye* (Oxford: Oxford University Press, 1982), pp. 40–62. There are two other important essays on expression in the same volume: 'The Mask and the Face: The Perception of Physiognomic Likeness in Life and in Art', pp. 105–30, and 'Ritualized Gesture and Expression in Art', pp. 63–77.

25 Edwin Clarke and L. S. Jacyna, *Nineteenth-Century Origins of Neuroscientific Concepts* (Berkeley: University of California Press, 1987), p. 102.

26 See Roger Smith, 'The Background of Physiological Psychology in Natural Philosophy', *History of Science* 11 (1973): 75–123, pp. 80–8. My discussion of physiological psychology (in the associationist form developed by Hartley) is indebted to Smith's extremely interesting and important work.

27 David Hartley, *Observations on Man, his Frame, his Duty, and his Expectations: In Two Parts* (London: S. Richardson, 1749), I, pp. i–ii.

28 Karl Figlio, 'Theories of Perception and the Physiology of Mind in the Late Eighteenth Century', *History of Science*, 12 (1975): 177–212, p. 177.

29 Hartley, *Observations on Man*, I, p. 12.

30 Ibid., I, pp. 62–3.

31 Compare this to Roger Smith's use of these two metaphors to describe the history of psychology: '[It] is not a history that details the discovery of law-like processes in the mind and pictures the path of scientific psychology. Rather, it is a history that shows the power of metaphor – a social metaphor, 'association', and a physical metaphor, 'ideas' – to create contested versions of scientific psychology.' *Fontana History of the Human Sciences* (London: Fontana, 1997), p. 251.

32 Robert J. Richards, *Darwin and the Emergence of Evolutionary Theories of Mind and Behaviour* (Chicago: University of Chicago Press, 1987), pp. 30–1.

33 Erasmus Darwin, *Zoonomia; or, the laws of organic life*, 2 vols. (Dublin: Byrne and Jones, 1794–6), I, pp. 524–5. I have borrowed this example from Robert M. Young, 'The Role of Psychology in the Nineteenth-Century Evolutionary Debate', *Historical Conceptions of Psychology*, ed. M. Henle, J. Jaynes, and J. J. Sullivan (New York: Springer, 1973), pp. 192–3.

34 Hartley, *Observations on Man*, II, pp. 512–13.

35 The literature on Lavater's famous work is extensive. For a selected list of references see notes 8–10 in the introduction.

36 References are to the Holcroft edition; see note 11.

37 Michael Shortland, '"Skin Deep": Barthes, Lavater, and the Legible Body', *Economy and Society* 14 (1985): 272–312, p. 285.

38 Lindley Darden, 'Character: Historical Perspectives', *Keywords in Evolutionary Biology*, ed. Evelyn Fox Keller and Elisabeth Lloyd (Cambridge, Mass.: Harvard University Press, 1992), p. 41. See also Kurt Fristrup's 'Character: Current Usages', pp. 45–51, in the same volume.

39 Lavater's use of silhouettes is interesting in this respect as they smooth out the detail of an individual's countenance and present a profile of its external line. There are many similarities between Lavater's use of the external line identified by the silhouette, Pieter Camper's construction of facial angles, and Francis Galton's composite portraits, particularly in terms of the supposed potential for identifying higher and lower orders of human beings and, in the case of Galton, selecting out the less desirable forms of appearance. For a more detailed discussion of Camper see chapter two; for more on Galton see chapter six.

40 Christopher Rivers, *Face Value: Physiognomical Thought and the Legible Body in Marivaux, Lavater, Balzac, Gautier, and Zola* (Madison: University of Wisconsin Press, 1994), p. 96.

41 [Anon.], 'The Projector, No. xc', *The Gentleman's Magazine* 78 (1808): p. 1085. Compare this to an earlier issue of the periodical where it is

stated that: 'a servant would, at one time, scarcely be hired till the descriptions and engravings of Lavater had been consulted, in careful comparison with the lines and features of the young man's or woman's countenance'. *The Gentleman's Magazine* 71 (1801): p. 124.

42 *The Encyclopaedia Britannica*, 8th edn, 22 vols. (Edinburgh: Archibald Constable & Co., 1853–60), XVIII, p. 576.

2 THE ARGUMENT FOR EXPRESSION: CHARLES BELL AND THE CONCEPT OF DESIGN

1 Charles Bell, 'Idea of a New Anatomy of the Brain', *Philosophical Transactions of the Royal Society of London* (1811), repr. in *The Way In and the Way Out: François Magendie, Charles Bell, and the Roots of the Spinal Nerves*, ed. Paul F. Cranefield (New York: Futura, 1974).

2 Charles Bell, *The Letters of Charles Bell*, ed. G. J. Bell (London: John Murray, 1870), p. 314.

3 Charles Bell, *The Hand: Its Mechanisms and Vital Endowments as Evincing Design* (London: William Pickering, 1833), 'Notice'.

4 On the treatises and their impact, see Jonathan Topham, 'An Infinite Variety of Arguments: The *Bridgewater Treatises* and British Natural Theology in the 1830s', PhD diss., University of Lancaster, 1993, and Richard Yeo, *Defining Science: William Whewell, Natural Knowledge, and Public Debate in Early Victorian Britain* (Cambridge: Cambridge University Press, 1993).

5 Bell, *The Hand*, p. 137.

6 Ludmilla Jordanova, 'The Representation of the Body: Art and Medicine in the Work of Charles Bell', *Towards a Modern Art World, Studies in British Art I*, ed. Brian Allen (New Haven, Conn: Yale University Press, 1995), pp. 80–1.

7 See *Sir Charles Bell, His Life and Times*, ed G. Gordon-Taylor and E. W. Walls (Edinburgh: E. & S. Livingstone, 1958), pp. 178–80.

8 In addition to the various editions of *System of Dissections*, Bell produced a further textbook on anatomy, *Engravings of the Arteries, of the Nerves and of the Brain* (1801–2); *Essays on the Anatomy of Expression* (1806); two papers for the Royal Society – 'Idea of a New Anatomy of the Brain' (1820) and 'On the Nerves of the Face' (1829); and three texts with natural theological themes – *The Hand: Its Mechanisms and Vital Endowments as Evincing Design* (1833), *Animal Mechanics, or, Proofs of Design in the Animal Frame* (1838), and *Familiar Treatises on the Five Senses* (1841) – along with a jointly produced edition with Henry, Lord Brougham, of Paley's *Natural Theology* (1836).

9 I am thinking here of Scottish moral philosophical writers such as Thomas Reid, Robert Whytt, Dugald Stewart, Thomas Brown – and slightly later James Mill – who were interested in the growth of physiological ideas within the sensualist tradition of John Locke, David

Hartley, and Joseph Hume, and the French school of Condillac, Tracy, Cabanis, Richerand, and Magendie See G. Bryson, *Man and Society: The Scottish Inquiry of the Eighteenth Century* (Princeton: Princeton University Press, 1945); and S. Grave, *The Scottish Philosophy of Common Sense* (Oxford: Oxford University Press, 1960).

10 See notes 1 and 3 for publication details.

11 This was first published as *Essays on the Anatomy of Expression* (London: George Bell & Sons, 1806) A second edition was published as *Essays on the Anatomy and Philosophy of Expression* (London: George Bell & Sons, 1824); and a substantially revised third edition was issued as *The Anatomy and Philosophy of Expression, as Connected with the Fine Arts* (London: George Bell & Sons, 1844). Enduring proof of the widespread appeal of Bell's book throughout the nineteenth century can be found in the subsequent editions published in 1847, 1865, 1873 (New York), 1877, and 1890. As the most extensive treatment of expression, I shall be referring to the revised third edition (1844) throughout this chapter.

12 L. S. Jacyna, 'Immanence or Transcendence: Theories of Life and Organization in Britain, 1790–1835', *Isis* (1983): 311–29, p. 312.

13 Roger Smith, *Inhibition: History and Meaning in the Sciences of Mind and Brain* (Berkeley: University of California Press, 1992), p. 41.

14 Ibid., p. 145.

15 Henry Holland, *Chapters on Mental Physiology* (London: Longman, Brown, Green and Longmans, 1852), p. 109; cited in Smith, *Inhibition*, p. 41.

16 Joseph Butler, *The Analogy of Religion Natural and Revealed to the Constitution and Course of Nature* (London, 1736), ii.iv, p. 4.

17 See Gillian Beer, *Darwin's Plots: Evolutionary Narrative in Darwin, George Eliot and Nineteenth-Century Fiction* (London: Ark Paperbacks, 1983), p. 84.

18 For an account of the emergence of natural theology up to and including Paley, see Charles C. Gillispie, *Genesis and Geology: A Study in the Relations of Scientific Thought, Natural Theology, and Social Opinion in Great Britain, 1790–1850* (Cambridge, Mass.: Harvard University Press, 1951, 1996), pp. 3–40.

19 William Paley, *Natural Theology; or, Evidences of the Existence and Attributes of the Deity collected from the Appearances of Nature* (1802), vol. I of *The Works of William Paley*, ed. G. W. Meadley, 5 vols. (Boston, 1810), I, p. 11.

20 Paley, *Natural Theology*, I, p. 16.

21 Gillispie, *Genesis and Geology*, p. 36.

22 Paley, *Natural Theology*, I, p. 367.

23 Jacyna, 'Immanence or Transcendence', p. 319.

24 Cuvier, cited by Adrian Desmond, *The Politics of Evolution: Morphology, Medicine, and Reform in Radical London* (Chicago: University of Chicago Press, 1989), pp. 111–12.

25 Desmond, *The Politics of Evolution*, p. 47. The following discussion is indebted to Coleman's and Desmond's histories of these debates.

W. Coleman, *Georges Cuvier, Zoologist. A Study in the History of Evolution Theory* (Cambridge, Mass.: Harvard University Press, 1964).

26 Cited by Coleman, *Georges Cuvier, Zoologist*, p. 42.

27 Cited by Coleman, *Georges Cuvier, Zoologist*, p. 39.

28 Desmond, *The Politics of Evolution*, pp. 48–9.

29 In addition to the impressive studies by Desmond and Coleman, see Toby Appel, *The Cuvier-Geoffroy Debate: French Biology in the Decades before Darwin* (New York: Oxford University Press, 1987); Dorinda Outram, *Georges Cuvier: Vocation, Science and Authority in Post-Revolutionary France* (Manchester: Manchester University Press, 1984); and Robert J. Richards, *Darwin and the Emergence of Evolutionary Theories of Mind and Behaviour* (Chicago: University of Chicago Press, 1987).

30 Desmond, *The Politics of Evolution*, p. 49.

31 See Desmond, *The Politics of Evolution*, pp. 92–100, for an explanation of Bell's place in the charged political climate of 1820s and 1830s scientific debate.

32 See chapter one for more on this.

33 Descartes described this as follows: 'the mind is not immediately affected by all parts of the body, but only the brain, or perhaps just by one small part of the brain, namely the part which is said to contain the "common" sense', 'Meditations on the First Philosophy', *The Philosophical Works of Descartes*, trans. John Cottingham, Robert Stoothoff, and Dugald Murdoch, 3 vols. (Cambridge: Cambridge University Press, 1985), II, p. 59. There is a large literature on this topic but one of the most comprehensive and sophisticated studies is Edwin Clarke and L. S. Jacyna's *Nineteenth-Century Origins of Neuroscientific Concepts* (Berkeley: University of California Press, 1987). There is also a related area of work which addresses the notion of the *sensus communis* in the light of Kant's presentation of it, especially in the third Critique. For this strand of thought, see Jay Bernstein, *The Fate of Art: Aesthetic Alienation from Kant to Derrida and Adorno* (Oxford: Polity Press, 1993); Howard Caygill, *The Art of Judgement* (Oxford: Basil Blackwell, 1989); David Summers, *The Judgement of Sense: Renaissance Naturalism and the Rise of Aesthetics* (Cambridge: Cambridge University Press, 1987).

34 'Idea of a New Anatomy of the Brain', p. 23.

35 As a result of his discovery, Bell became embroiled in a heated dispute with François Magendie over who was the first to correctly identify the function of the anterior and the posterior roots. Despite a huge row about this, it is now held that Bell presented the first, preliminary findings but Magendie is credited with correctly identifying the functional bases of both roots and convincingly demonstrating the claim. For more on the disagreement between Bell and Magendie, see Paul Cranefield, *The Way in and the Way Out: François Magendie, Charles Bell and the Roots of the Spinal Nerves* (New York: Futura, 1974); and Clarke and Jacyna, *Nineteenth-Century Origins of Neuroscientific Concepts*, pp. 110–14.

36 John and Charles Bell, *The Anatomy of the Human Body*, 4 vols. (London, 1793–1804), III, p. 38.

37 Karl Figlio, 'Theories of Perception and the Physiology of Mind in the Late Eighteenth Century', *History of Science* 12 (1975): 177–212, p. 177.

38 Cited by Figlio, 'Theories of Perception', p. 184.

39 Figlio, 'Theories of Perception', p. 178.

40 Bell, 'Idea of a New Anatomy of the Brain', pp. 106–7.

41 Ibid., p. 105. The quotation reads in full: 'The prevailing doctrine of the anatomical schools is, that the whole brain is a common sensorium; that the extremities of the nerves are organized, so that each is fitted to receive a peculiar impression; or that they are distinguished from each other only by delicacy of structure, and by a corresponding delicacy of sensation; that the nerve of the eye, for example, differs from the nerves of touch only in the degree of its sensibility.'

42 Bell, 'Idea of a New Anatomy of the Brain', pp. 107–8.

43 Ibid., p. 110.

44 Bell, *The Hand*, p. 255.

45 Ibid., p. 214.

46 Gillian Beer, 'Four Bodies on the Beagle: Touch, Sight, and Writing in a Darwin Letter', *Textuality and Sexuality: Reading Theories and Practices*, ed. Judith Still and Michael Worton (Manchester: Manchester University Press, 1993), pp. 119–20.

47 Bell, *The Hand*, pp. 148–9.

48 Ibid., p. 190.

49 Ibid., p. 211.

50 Page references are to the revised third edition detailed in note 11.

51 Leonardo da Vinci is the obvious example of an artist who applied the techniques of anatomy to painting and architectural design. For more on the connection of art and anatomy, see William F. Bynum and Roy Porter, eds., *Medicine and the Five Senses* (Cambridge: Cambridge University Press, 1993); Martin Kemp, *The Science of Art: Optical Themes in Western Art from Brunelleschi to Seurat* (New Haven: Yale University Press, 1990); and Jonathan Sawday, *The Body Emblazoned: Dissection and the Human Body in Renaissance Culture* (London: Routledge, 1995).

52 For more detail on medicine as an education of the senses, see Ludmilla Jordanova, 'Body Image and Sex Roles', *Sexual Visions: Images of Gender in Science and Medicine between the Eighteenth and Twentieth Centuries* (Hemel Hempstead: Harvester Wheatsheaf, 1989), pp. 43–65; Sander Gilman, 'Touch, Sexuality and Disease'; and Roy Porter, 'The Rise of Physical Examination', *Medicine and the Five Senses*, ed. William F. Bynum and Roy Porter (Cambridge: Cambridge University Press, 1993), pp. 198–224 and pp. 170–97 respectively.

53 Benjamin Robert Haydon, *The Autobiography and Journals of Benjamin Robert Haydon*, ed. Malcolm Elwin (London: Macdonald, 1950), pp. 37–8.

54 Ibid., p. 36.
55 Ibid., p. 36.
56 Ibid., p. 34.
57 Ibid., p. 122.
58 Transcribed by A. N. L. Munby, 'The Bibliophile: B. R. Haydon's Anatomy Book', *Apollo* 26 (1937): 345–7, p. 345.
59 Cited by Ilaria Bignamini and Martin Postle, *The Artist's Model: Its Role in British Art from Lely to Etty* (Nottingham: Nottingham University Art Gallery, 1991), p. 32.
60 On this theme, see Frederick Cummings, 'Charles Bell and *The Anatomy of Expression*', *The Art Bulletin* 46 (1964): 191–203.
61 Hearing that the Queen spent two hours one evening reading his work, Bell is reputed to have remarked nonchalantly: 'Oh happiness in the extreme, that I should ever write anything fit to be dirtied by her snuffy fingers'. See Gordon-Taylor and Walls, eds., *Sir Charles Bell. His Life and Times*, p. 21.

3 WHAT IS CHARACTER? THE NATURE OF ORDINARINESS IN THE PAINTINGS OF THE PRE-RAPHAELITE BROTHERHOOD

1 Hereafter, I shall refer to the Pre-Raphaelite Brotherhood in its commonly abbreviated form of 'PRB'.
2 Bell, *Anatomy of Expression*, p. 194.
3 William Michael Rossetti, 'The Royal Academy Exhibition, 1861', *Fraser's Magazine* (1861): p. 62.
4 William Holman Hunt, *Pre-Raphaelitism and the Pre-Raphaelite Brotherhood*, 2 vols. (London: Macmillan & Co., 1905), I, p. 48.
5 William Michael Rossetti, ed., *Dante Gabriel Rossetti: His Family Letters, with a Memoir*, 2 vols. (London, 1895), I, p. 135.
6 See William Michael Rossetti, *The P. R. B. Journal: William Michael Rossetti's Diary of the Pre-Raphaelite Brotherhood, 1849–1853*, ed. William E. Fredeman (Oxford: Clarendon Press, 1975), p. 96. In one of his letters to his brother Dante Gabriel wrote playfully: 'Hunt and I have prepared a list of Immortals, forming our creed, and to be pasted up on our study for the affixing of all decent fellows' signatures. It has already caused considerable horror among our acquaintance. I suppose we shall have to keep a hairbrush.' Oswald Doughty and J. R. Wahl, eds., *Letters of Dante Gabriel Rossetti*, 4 vols. (Oxford: Clarendon Press, 1965–7), I, p. 35.
7 Bell, *Anatomy of Expression*, p. 18.
8 Ibid.
9 See Charlotte Klonk, *Science and the Perception of Nature: British Landscape Art in the late Eighteenth and Early Nineteenth Century* (New Haven: Yale University Press, 1996).
10 Hunt, *Pre-Raphaelitism*, I, p. 101.

11 The literature on the PRB, the artists and their work, is extensive, but see in particular the following works: Hilary Fraser, *Beauty and Belief: Aesthetics and Religion in Victorian Literature* (Cambridge: Cambridge University Press, 1986); William E. Fredeman, *Pre-Raphaelitism: A Bibliocritical Study* (Cambridge, Mass.: Harvard University Press, 1965); Timothy Hilton, *The Pre-Raphaelites* (London: Thames and Hudson, 1970); John Dixon Hunt, *The Pre-Raphaelite Imagination: 1848–1900* (London: Routledge & Kegan Paul, 1968); George Landow, *William Holman Hunt and Typological Symbolism* (New Haven: Yale University Press, 1979); Leslie Parris, ed., *Pre-Raphaelite Papers* (London: The Tate Gallery, 1984); Marcia Pointon, ed., *Pre-Raphaelites Re-Viewed* (Manchester: Manchester University Press, 1989); James Sambrook, *Pre-Raphaelitism: A Collection of Critical Essays* (Chicago: University of Chicago Press, 1974); Lindsay Smith, *Victorian Photography, Painting, and Poetry: The Enigma of Visibility in Ruskin, Morris and the Pre-Raphaelites* (Cambridge: Cambridge University Press, 1995). The best range of Pre-Raphaelite art is contained in Alan Bowness' edited catalogue, *The Pre-Raphaelites* (London: The Tate Gallery and Penguin, 1984).

12 Hunt, *Pre-Raphaelitism*, I, p. 48. It is interesting that Hunt conflates the fashion for waxwork figures in the late eighteenth and early nineteenth-centuries – which was affirmed by the contemporaneous opening of Madame Tussauds – with the constrained poses of the models of the Royal Academy life class.

13 Ibid., p. 51.

14 Dante Gabriel Rossetti, 'The Portrait', in 'The House of Life' (1868), *The Poetical Works of Dante Gabriel Rossetti*, ed. William Michael Rossetti (London: Elvis and Elvey, 1900), p. 78.

15 John Keats, *Poetical Works*, ed. H. W. Garrod (Oxford: Oxford University Press, 1956), pp. 179 and 184. For details of the sometimes quite controversial reception of Keats' poetry, see G. H. Ford, *Keats and the Victorians: A Study of the Influence and Rise to Fame, 1821–1895* (London: Yale University Press, 1944).

16 For a description of the Hogarthian qualities also found in Hunt's work, see Landow, *William Holman Hunt and Typological Symbolism*, pp. 38–9 and 47–59.

17 William Michael Rossetti, *Fine Arts, Chiefly Contemporary* (London: Macmillan & Co., 1867), p. 187.

18 [John Tupper], 'The Subject in Art (No. 1)', *The Germ: Thoughts Towards Nature in Poetry, Literature, and Art* I (January 1850): 11–18, p. 14 (repr. Oxford: Ashmolean Museum, 1992).

19 Ibid., p. 15.

20 [Tupper], 'The Subject in Art (No. 2)', *The Germ* 3 (March 1850): 118–25, p. 122. A few pages earlier, Tupper invokes an image of the city in order to distinguish active from passive poets, that is to say, those that see and feel from those that see but do not feel. He writes: 'For let a

poet walk through London, and he shall see a succession of incidents, suggesting some moral beauty by a contrast of times with times, unfolding some principle of nature, developing some attribute of man, or pointing to some glory in the Maker: while the man who walked behind him saw nothing but shops and pavements, and coats and faces; neither did he hear the aggregated turmoil of a city of nations, nor the noisy exponents of various desires, appetites and pursuits: each pulsing tremor of the atmosphere was not struck into it by a subtle [*sic*] ineffable something willed forcibly out of a cranium (p. 120).

21 John Seward (Frederick George Stephens), 'The Purpose and Tendency of Early Italian Art', *The Germ* 2 (February 1850): 58–64, p. 61.

22 [Anon.], *The Times* (6 May 1851). Hunt quotes this review in *Pre-Raphaelitism*, 1, p. 249.

23 There is an invaluable collection of Victorian critics' responses to contemporary art edited by John C. Olsmstead, *Victorian Painting: Essays and Reviews*, 2 vols. (New York: Garland Publishing, 1980). For a discussion of the reception of PRB paintings, see Herbert Sussman, 'The Language of Criticism and the Language of Art: The Responses of Victorian Periodicals to the Pre-Raphaelite Brotherhood', *Victorian Periodicals Newsletter* 19 (1979): 15–29.

24 For instance, Stephanie Grilli discusses the assimilation of medical discourse into the art-critical language used to attack the PRB paintings. See 'Pre-Raphaelite Portraiture, 1848–1854', PhD diss. (Yale University, 1980), pp. 5–10.

25 Julie F. Codell, 'Expression over Beauty: Facial Expression, Body Language, and Circumstantiality in the Paintings of the Pre-Raphaelite Brotherhood', *Victorian Studies* 29 (1986): 255–90.

26 Ralph N. Wornum, 'Modern Moves in Art: Christian Architecture. Young England', *The Art-Journal* 12 (1850): 117–18.

27 See most notably, [Anon.], 'Royal Academy', *Athenaeum* (1850): 508–9, 533–5, 558–60, 590–1 & 615–17; Charles Dickens, 'Old Lamps for New Ones', *Household Words: A Weekly Journal, conducted by Charles Dickens*, I (1850): 265–7; and David Masson, 'Pre-Raphaelitism in Art and Literature', *British Quarterly Review* 16 (1852): 197–220.

28 Mosche Barasch, *Giotto and the Language of Gesture* (Cambridge: Cambridge University Press, 1981), p. 3.

29 Barasch, *Giotto and the Language of Gesture*, pp. 3–4. On the same theme, see also Richard Brilliant, *Gesture and Rank in Roman Art* (New Haven: Yale University Press, 1983).

30 Hunt, *Pre-Raphaelitism*, 1, p. 85.

31 Ibid., pp. 85–6.

32 Ibid., p. 98.

33 Millais' work was submitted to the Royal Academy for its 1848 exhibition but was rejected because, according to Hunt, it was unfinished. It was however bought in 1849 by James Wyatt, an Oxford

collector and dealer, for £60, whilst Hunt's accepted work was chosen somewhat reluctantly by Charles Bridger, winner of the Art-Union's lottery prize in 1860.

34 Keats, 'The Eve of St. Agnes', *Poetical Works*, p. 206.

35 In Bowness, *The Pre-Raphaelites*, p. 57.

36 Lindsay Smith, *Victorian Photography, Painting and Poetry: The Enigma of Visibility in Ruskin, Morris, and the Pre-Raphaelites* (Cambridge: Cambridge University Press, 1995), p. 93.

37 Ibid., p. 96.

38 Ibid., p. 96.

39 John Ruskin, *Modern Painters, The Collected Works of John Ruskin*, ed. E. T. Cook and Alexander Wedderburn, 39 vols. (London: George Allen, 1903–12), v, p. 99. Interestingly, Ruskin mentions Bell's *Anatomy of Expression* twice in the previous volume of *Modern Painters*; see *Collected Works*, iv, pp. 159 and 179 respectively.

40 Ruskin, *The Times* (13 May 1851); repr. Robert L. Hewison, ed., *The Art Criticism of John Ruskin* (New York: Da Capo Press, 1964), p. 373.

41 Ruskin, *The Times* (13 May 1853); repr. Hewison, ed., *Art Criticism*, pp. 383–4.

42 Joshua Reynolds, 'Discourse iv', *Discourses on Art*, ed. Robert R. Wark (New Haven: Yale University Press, 1975), pp. 58–61.

43 The classic account of the function of painting in the commercial society of the late eighteenth century is John Barrell's *The Political Theory of Painting from Reynolds to Hazlitt – 'The Body of the Public'* (New Haven: Yale University Press, 1986).

44 Reynolds, 'Discourse i', *Discourses on Art*, p. 13.

45 Ibid., pp. 60–1.

46 There is a large literature on Ruskin but as yet there is no comprehensive study of his intellectual debt to Reynolds. A number of studies are notable, however, for addressing aspects of the eighteenth-century tradition out of which Ruskin emerges; see, for example, Patricia Ball, *The Science of Aspects: The Changing Role of Fact in Coleridge, Ruskin, and Hopkins* (London: The Athlone Press, 1971); Linda Dowling, *The Vulgarization of Art: The Victorians and Aesthetic Democracy* (Charlottesville: University of Virginia Press, 1996); Elisabeth Helsinger, *Ruskin and the Art of the Beholder* (Cambridge, Mass.: Harvard University Press, 1982); Robert Hewison, *John Ruskin, The Argument of the Eye* (Princeton: Princeton University Press, 1976); Martin Meisel, *Realizations: Narrative, Pictorial, Theatrical Arts in Nineteenth-Century England* (Princeton: Princeton University Press, 1983); Richard Stein, *The Ritual of Interpretation: The Fine Arts as Literature in Ruskin, Rossetti, and Pater* (Cambridge, Mass.: Harvard University Press, 1975).

47 For specific studies of Brown which look at his understanding of history painting, see Ford Madox Ford (Heuffer), *Ford Madox Brown: A Record of his Life and Work* (London: Longman's Green, 1896); and Lucy Rabin,

Ford Madox Brown and the Pre-Raphaelite History-Picture (New York: Garland Publishing, 1978).

48 Ford Madox Brown, 'On the Mechanism of an Historical Picture. Part I – The Design', *The Germ* 2 (February 1850): 60–3, p. 60.

49 Michael Fried, *Absorption and Theatricality: Painting and the Beholder in the Age of Diderot* (Chicago: University of Chicago Press, 1980).

50 Ibid., p. 32.

51 Ibid., p. 75.

52 Ibid., p. 61.

53 Ibid., pp. 131–2.

54 Henry Siddons, *Practical Illustrations of Rhetorical Gesture and Action; adapted to the English Drama, from a work on the subject by M. Engel, member of the Royal Academy of Berlin* (1807), 2nd edn (London: Sherwood, Neely & Jones, 1822).

55 Ibid., pp. 17–18.

56 Ibid., p. 279.

57 Ibid., p. 27.

58 Ibid., p. 21.

59 George Grant, *An Essay on the Science of Acting* (London: Cowie & Strange, 1828).

60 Ibid., pp. 6–7.

61 Eliza Lynn, 'Passing Faces', *Household Words: A Weekly Journal conducted by Charles Dickens*, 11 (1855): p. 261.

62 Ibid., p. 262.

63 Ibid., p. 261.

64 Ibid., p. 264.

65 Elizabeth Eastlake, [Review article], *Quarterly Review* 90 (December 1851): 62–91, p. 62.

66 Ibid., pp. 62–3.

67 Ibid., p. 72.

68 Ibid., p. 71.

69 Thomas Woolnoth, *Facts and Faces: Linear and Mental Portraiture morally considered, and pictorially illustrated* (London, 1852).

70 Ibid., p. 22.

71 Ford Madox Brown, *The Exhibition of Work and Other Paintings* (London: McCorquodale & Co., 1865), p. 26.

72 Ibid., p. 8.

73 Ibid., p. 8.

74 See Mary Cowling, *The Artist as Anthropologist: The Representation of Type and Character in Victorian Art* (Cambridge: Cambridge University Press, 1989).

75 [Anon.], 'The Physiognomy of the Human Form', *Quarterly Review* 99 (1856): 532–91, pp. 457–61.

76 Ibid., pp. 457–8.

77 Ibid., p. 461.

4 'BEAUTY OF CHARACTER AND BEAUTY OF ASPECT':
EXPRESSION, FEELING, AND THE CONTEMPLATION OF EMOTION

1 Alexander Bain, *The Senses and the Intellect* (London: John W. Parker and Son, 1855), p. 336.
2 Herbert Spencer, 'The Haythorne Papers No. VIII. Personal Beauty', *Leader* 5 (1854): 356–7, p. 356.
3 Herbert Spencer, *An Autobiography*, 2 vols. (London: Williams and Norgate, 1904), I, p. 478. The context of these remarks makes for interesting reading. Writing to friends in 1856, Spencer rejected their suggestion that marriage would be a cure for his physical ailments and mental exhaustion. He wrote: 'I am perfectly willing to try your remedy for rationalism. Indeed, marriage has been prescribed as a means of setting my brain right in quite another sense: the companionship of a wife being considered the best distraction – in the French not the English meaning of the word. But the advice is difficult to follow. I labour under the double difficulty that my choice is very limited and that I am not easy to please. Moral and intellectual beauties do not by themselves suffice to attract me; and owing to the stupidity of our educational system it is rare to find them united to a good physique. Moreover, there is the pecuniary difficulty. Literature, and especially philosophical literature, pays badly. If I married, I should soon have to kill myself to get a living.'
4 Herbert Spencer, 'Physical Training', *British Quarterly Review* (1858–9), p. 395.
5 Herbert Spencer, 'The Haythorne Papers No. IX. Personal Beauty – Part II', *Leader* 5 (1854): 451–2, p. 452.
6 Spencer, 'Personal Beauty', p. 357.
7 See Peter J. Bowler, 'Herbert Spencer and "Evolution": An Additional Note', *Journal of the History of Ideas*, 36 (1975): p. 367; John W. Burrow, *Evolution and Society: A Study in Victorian Social Thought* (Cambridge: Cambridge University Press, 1966); J. D. Y. Peel, *Herbert Spencer: The Evolution of a Sociologist* (London: Heinemann, 1971); and R. Richards, *Darwin and the Emergence of Evolutionary Theories of Mind and Behaviour* (Chicago: University of Chicago Press, 1987).
8 Spencer, 'Personal Beauty', p. 357.
9 Nancy L. Paxton, *George Eliot and Herbert Spencer: Feminism, Evolutionism, and the Reconstruction of Gender* (Princeton: Princeton University Press, 1991), p. 33.
10 Sander Gilman, *Health and Illness: Images of Difference* (London: Reaktion Books, 1995), p. 92.
11 Herbert Spencer, *The Principles of Psychology* (London: Longman, Brown, Green, and Longmans, 1855), pp. 600–1.
12 Spencer, 'Personal Beauty–Part II', p. 451.
13 Ibid.

14 For an idea of the extent of such writings see Jenny Bourne Taylor and Sally Shuttleworth, *Embodied Selves: An Anthology of Psychological Texts, 1830–1890* (Oxford: Clarendon Press, 1998).

15 Robyn Cooper, 'Victorian Discourses on Women and Beauty: The Alexander Walker Texts', *Gender and History* 5 (1993): 34–55, p. 34. See also Cooper, 'Definition and Control: Alexander Walker's Trilogy on Woman', *History of Sexuality* 2 (1992): 341–64.

16 Cooper, 'Victorian Discourses on Women and Beauty', p. 35.

17 Alexander Walker was a middle-class radical and Lamarckian who published a series of books on beauty and femininity. Notable amongst his works is an early treatise on physiognomy which applies physiological understanding to Lavaterian principles. See *Physiognomy founded on Physiology and applied to Various Countries* (London: Smith & Elder, 1834).

18 Alexander Walker, *Beauty: Illustrated Chiefly by an Analysis and Classification of Beauty in Woman. Preceded by a Critical View of the General Hypotheses respecting Beauty, by Hume, Hogarth, Burke, Knight, Alison, etc. and followed by a similar view of the hypotheses of beauty in sculpture and painting, by Leonardo da Vinci, Winckelmann, Mengs, Bossi, etc.* (London: Henry G. Bohn, 1837). The other parts of this trilogy develop similar, and by now familiar, themes. See *Intermarriage: or, the mode in which, and the causes why, beauty, health and intellect result from certain unions, and deformity, disease and insanity from others* (London: A. H. Baily & Co., 1838); and *Woman, Physiologically considered as to Mind, Morals, Marriage, Matrimonial Slavery, Infidelity and Divorce* (London: A. H. Baily & Co., 1839).

19 Ibid., p. vii.

20 Ibid., pp. vii-viii.

21 Ibid., p. 148.

22 Ibid., pp. viii-ix.

23 Ibid., p. xiv.

24 In this context, it is worth noting Robyn Cooper's claim that Walker's work is 'taken directly and without acknowledgment' from T. Bell's 1821 work on beauty, *Kalogynomia or the Laws of Female Beauty: being the Elementary Principles of that Science* (London: Walpole Press, 1821).

25 Walker, *Beauty*, pp. 12–13.

26 Ibid., p. 20. In a longer version of this passage, Walker expands on his sense of the connection between art and civilization. He wrote: 'A just sense of this truth [the critical judgement and pure taste for beauty] will give high encouragement to sculpture and painting–arts which may everywhere be looked upon as the best tests, as well as the best records, of civilization. Such encouragement they need in truth; for the monstrous monopoly of landed property and the accumulation of wealth in few hands – the great aim of our political economy, renders art poor indeed . . . A diffusion of wealth alone can give encouragement to art; nor can this ever be, while British industry is crushed under the weight of enormous taxation' (p. 22).

27 Walker, *Beauty*, p. 147.

28 Ibid., p. 161.

29 Walker, *Woman*, p.13.

30 For a lucid survey of the meaning and value of particular antique statues, like the Venus de Medici, in different historical periods, see Francis Haskell and Nicholas Penny, *Taste and the Antique: The Lure of Classical Sculpture, 1500–1900* (New Haven: Yale University Press, 1981).

31 See Alex Potts, *Flesh and the Ideal: Winckelmann and the Origins of Art History* (New Haven: Yale University Press, 1994), p. 2.

32 Walker, *Beauty*, pp. 365–6.

33 Ibid., pp. 3–4.

34 Cooper, 'Victorian Discourses on Women and Beauty', p. 38. In particular, Cooper draws attention to the furore over the exhibition of William Etty's *Ulysses and the Sirens* (1838) in Manchester. Public feeling was such that the painting was hung facing the wall. For more on the controversy surrounding this picture, see Stuart Macdonald, 'The Royal Manchester Institution', *Art and Architecture in Victorian Manchester*, ed. John H. G. Archer (Manchester: Manchester University Press, 1985).

35 Walker, *Beauty*, p. 18.

36 John Barrell gives an excellent analysis of the role of the figure of Venus in civic humanist discourses. Arguing that the notion of a civic character may not exclude the possibility of aesthetic pleasure, Barrell suggests that 'the claim made by the civic discourse, that it is possible to subtract the sensual from the aesthetic, or to detach the aesthetic from the sensual, and so to enjoy Venus's body on aesthetic terms while remaining unmoved by her sensuality, may have come to serve some new purposes'. See *The Birth of Pandora and the Division of Knowledge* (Houndsmills: Macmillan Press, 1992), p.83.

37 Herbert Marcuse, 'The Affirmative Character of Culture', *Negations: Essays in Critical Theory*, trans. J. L. Shapiro (Boston, Mass.: Beacon, 1968), p. 114. Explaining the need for beauty to be understood as art in the sphere of bourgeois culture, Marcuse argued that 'beauty is fundamentally shameless. It displays what may not be promised openly and what is denied the majority' (p.115).

38 Marcuse, 'Affirmative Character of Culture', p.119. Marcuse writes: 'If the individual is ever to come under the power of the ideal to the extent of believing that his concrete longings and needs are to be found in it – found moreover in a state of fulfillment and gratification, then the ideal must give the illusion of granting present satisfaction. It is this illusory reality that neither philosophy nor religion can attain. Only art achieves it – in the medium of beauty.'

39 Lynda Nead, *The Female Nude: Art, Obscenity and Sexuality* (London: Routledge, 1992), p. 7. On this theme, see also Casey Finch, ' "Hooked and Unbuttoned Together": Victorian Underwear and Representations

of the Female Body', *Victorian Studies* 34 (1991): 340–58; and Marcia Pointon, *Naked Authority: The Body in Western Painting 1830–1908* (Cambridge: Cambridge University Press, 1990).

40 John G. MacVicar, *On the Beautiful, the Picturesque, the Sublime* (London: Scott, Webster & Geary, 1837), p. 19.

41 Ibid., pp. 37–8.

42 Ibid., p. 35.

43 Spencer, *Principles of Psychology*, pp. 585–6.

44 Bain's study of character makes interesting (if somewhat idiosyncratic) reading because, like John Stuart Mill, Bain argued for the development of a science of character based on laws derived from the laws of mind. See Alexander Bain, 'On the Study of Character, including an Estimate of Phrenology' (London, 1861). For details of this synthesis and its place in the history of psychology, see Robert M. Young, 'The Role of Psychology in the Nineteenth-Century Evolutionary Debate', *Historical Conceptions of Psychology*, ed. M. Henle, J. Jaynes & J. J. Sullivan (New York: Springer Publishing, 1973), pp. 180–204.

45 Alexander Bain, *The Emotions and the Will*, 2nd edn (London, 1865), pp. 435–7. For the wider context and logic of this claim, see L. S. Jacyna, 'The Physiology of Mind, the Unity of Nature, and the Moral Order in Victorian Thought', *British Journal for the History of Science* 14 (1981): 109–32.

46 Bain, *The Senses and the Intellect*, p. 1.

47 The terms are Bain's own. See *The Senses and the Intellect*, p. 2.

48 Ibid.

49 The philosophical literature on this issue is immense. The classic statement of what subjectivity means is Thomas Nagel's 'What is it like to be a Bat?', *Mortal Questions*, ed. Thomas Nagel (Cambridge: Cambridge University Press, 1979), pp. 165–81. See also John Searle, *The Rediscovery of the Mind* (Cambridge, Mass.: MIT Press, 1992). Following Searle, some philosophers have adopted the term 'qualia' to describe the subjective quality of emotion. For an analysis of the controversies surrounding the subjective status of emotion, see David Pugmire, *Rediscovering Emotion* (Edinburgh: Edinburgh University Press, 1998).

50 Bain, *The Senses and the Intellect*, p. 84n.

51 Ibid., p. 88.

52 Wilkie Collins, *The Woman in White*, ed. Harvey Sucksmith (Oxford: World's Classics, 1973), p. 40. Hereafter, page references in the text will be to this edition.

53 [Anon.], *Saturday Review* 10 (1860): 249–50.

54 [Anon.], 'Our Female Sensation Novelists', *Our Living Age* 78 (1863): 352–69, p. 354. Whilst this review essay talks about sensation in generic terms, it concentrates in particular on Mrs Norton's *Lost and Saved* (1862), Mrs Henry Wood's *East Lynne* (1862) and *Verner's Pride* (1863), and Elizabeth Braddon's *Lady Audley's Secret* (1862) and *Aurora Floyd* (1863).

55 See Jeanne Fahnestock, 'The Heroine of Irregular Features: Physiognomy and the Conventions of Heroine Description', *Victorian Studies* 24 (1981): 325–30. On this theme, see also: Graeme Tytler, *Physiognomy in the European Novel: Faces and Fortunes* (Princeton: Princeton University Press, 1982); Julie McMaster, *The Index of the Mind: Physiognomy in the Novel* (Lethbridge: University of Lethbridge Press, 1990); and Fanny Price, 'Imagining Faces: The Later Eighteenth-Century Sentimental Heroine and the Legible, Universal Language of Physiognomy', *British Journal of Eighteenth-Century Studies* 6 (1983): 1–16.

56 D. A. Miller, ' "Cage aux folles": Sensation and Gender in Wilkie Collins' *The Woman in White*', *Representations* 14 (1986): 107–36, pp. 107–8.

57 Ibid., p. 108.

58 Ann Cvetkovich, *Mixed Feelings: Feminism, Mass Culture, and Victorian Sensationalism* (New Brunswick: Rutgers University Press, 1992), p. 23.

59 Ibid., p. 17.

60 Cvetkovich is thoroughly in tune with Marxist literary critics on the potential of sensation narratives to reveal social contradictions. See, for example, Jonathan Loesberg, 'The Ideology of Narrative Form in Sensation Fiction', *Representations* 13 (1986): 115–38; and Walter M. Kendrick, 'The Sensationalism of *The Woman in White*', *Nineteenth-Century Fiction* 32 (1977): 18–35.

61 The phrase appears in an important passage for the anatomy of sensation as Basil details the stage-by-stage process of his delirium. Wilkie Collins, *Basil. A Story of Modern Life* (New York: Dover Publications, 1980). Subsequent references are to this edition.

62 Collins, *Basil*, pp. iii–v. Collins gives a more light-hearted account of the creative process in a subsequent response to a reader: 'I have got my idea; I have got three of my characters. What is there to do now? My next proceeding is to begin building up the story. Here my favourite three efforts must be encountered. First effort: To begin at the beginning. Second effort: To keep the story always advancing, without paying the smallest attention to the serial division in parts, or to the book publication in volumes. Third effort: To decide on the end.' 'How I Write my Books: Related in a Letter to a Friend', *The Globe* (1887): 6.

63 For more on this theme, see Ludmilla Jordanova, 'Nature Unveiling Before Science', *Sexual Visions: Images of Gender in Science and Medicine between the Eighteenth and Twentieth Centuries* (Hemel Hempstead: Harvester Wheatsheaf, 1989), pp. 87–110.

64 Believing himself to have disposed of Mannion in a gruesome encounter, the enormity of Basil's actions suddenly dawns: ' "MAD!" – that word, as I heard it, rang after me like a voice of judgement. "MAD" – a fear had come over me, which, in all its frightful complication, was expressed by that one word – a fear which, to the man who suffers by it, is worse even than the fear of death; which no human language ever conveyed, or ever will convey, in all its horrible reality, to others' (p. 165).

65 Wilkie Collins, *No Name*, ed. Virginia Blain (Oxford: World's Classics, 1986), p. 146.

66 For an evaluation of the connection of psychological thought with artistic and literary concerns, see in particular Ekbart Faas, *Retreat into the Mind: Victorian Poetry and the Rise of Psychiatry* (Princeton: Princeton University Press, 1988); and Jenny Bourne Taylor, *In the Secret of the Home: Wilkie Collins, Sensation Narrative, and Nineteenth-Century Psychology* (London: Routledge, 1988).

67 Collins, *No Name*, p. 13.

68 [Margaret Oliphant], '[Review of] Wilkie Collins' *No Name*', *Blackwood's Magazine* 94 (1863): 170.

69 [Margaret Oliphant], 'Sensation Novels', *Blackwood's Magazine* 91 (1862): 564–84, p. 568. This is an earlier and slightly more complimentary review, but Oliphant's reservations are explicit: 'The rise of a Sensation School of art in any department is a thing to be watched with jealous eyes; but nowhere is it so dangerous as in fiction, where the artist cannot resort to a daring physical plunge, as on the stage, or to a blaze of palpable colour, as in the picture-gallery, but must take the passions and emotions of life to make his effects withal.'

70 [Oliphant], 'Sensation Novels', p. 572.

71 [Anon.], 'Our Female Sensation Novelists', p. 352.

72 Rev. W. T. Clarke, *The Phrenological Miscellany: or, The Annuals of Phrenology and Physiognomy, from 1865 to 1873* (New York: Fowler & Wells Co., 1887), p. 325.

73 Ibid.

74 Ibid., p. 326.

75 Ibid., p. 327.

76 Ibid., p. 328. A remarkable anaphoric sequence is used to affirm and validate Clarke's conviction. He explains: '*To be beautiful*, we must feed the spark of intellectual fire by reading and meditation, until it burns in steady flame, irradiating the face by its brilliancy, suffusing the countenance with light. *To be beautiful*, we must put a great, organizing, and ennobling purpose into the will, and concentrate our thought and affection upon it until enthusiasm wells up in the heart, and suffuses the countenance, and rebuilds the body on its own divine plan. *To be beautiful*, we must cherish every kind impulse and generous disposition, making love the ruling affection of the heart and the ordering principle and inspiring motive of life' (p. 328, my italics).

5 UNIVERSAL EXPRESSIONS: DARWIN AND THE
NATURALISATION OF EMOTION

1 Herbert Spencer, *The Principles of Psychology* (London: 1855), p. 596.

2 See M. J. S. Hodge and D. Kohn, 'The Immediate Origins of Natural

Selection', *The Darwinian Heritage*, ed. David Kohn (Princeton: Princeton University Press, 1986), pp. 185–206.

3 Charles Darwin, *The Descent of Man, and Selection in relation to Sex* (London: John Murray, 1871); repr. *The Descent of Man, and Selection in relation to Sex* (Princeton: Princeton University Press, 1981). References in this chapter are to this reprinted edition.

4 Charles Darwin, *The Expression of the Emotions in Man and Animals* (London: John Murray, 1872). Until recently the only edition available was a reprint of this first edition, introduced by Konrad Lorenz: *The Expression of the Emotions in Man and Animals* (Chicago: University of Chicago Press, 1965). A new edition by Paul Ekman was published as I was completing the final revisions for this book. Ekman's edition, which he calls the third edition, contains the revisions Darwin had made but were not included in the second edition edited by his son (1889); see *The Expression of the Emotions in Man and Animals. Definitive Edition*, ed. Paul Ekman (London: HarperCollins, 1998). Ekman's edition contains commentaries on the text, a number of appendices mainly on the photographs and illustrations, and an afterword which brings Darwin's research up to date. The format of Ekman's edition is, however, diverting as he interpolates his own commentary within the body of Darwin's text. References in this chapter are to Darwin's first edition.

5 Charles Darwin, *On the Origin of Species by Means of Natural Selection, or The Preservation of Favoured Races in the Struggle for Life* (London: John Murray, 1859); repr. *The Origin of Species*, ed. Gillian Beer (Oxford: World's Classics, 1997). References in this chapter are to this edition.

6 The literature on Darwinian thought, and in particular the evolution of man, is extensive. Those works with most relevance to this study are the following: Gillian Beer, *Darwin's Plots: Evolutionary Narrative in Darwin, George Eliot and Nineteenth-Century Fiction* (London: Ark, 1983); Peter J. Bowler, *Biology and Social Thought, 1850–1914* (Berkeley: University of California & Office for History of Science & Technology, 1993); Michael T. Ghiselin, *The Triumph of the Darwinian Method* (Berkeley: University of California Press, 1969); John C. Greene, *The Death of Adam: Evolution and its Impact on Western Thought* (New York: Mentor, 1961); David Kohn, ed., *The Darwinian Heritage* (Princeton: Princeton University Press, 1985); James Moore, ed., *History, Humanity, and Evolution: Essays for John C. Greene* (Cambridge: Cambridge University Press, 1989); David Oldroyd and Ian Langham, eds., *The Wider Domain of Evolutionary Thought* (Dordrecht: Reidel, 1983); Robert J. Richards, *Darwin and the Emergence of Evolutionary Theories of Mind and Behaviour* (Chicago: University of Chicago Press, 1987); Frank M. Turner, *Between Science and Religion: The Reaction to Scientific Naturalism in Late Victorian England* (New Haven: Yale University Press, 1974); Robert M. Young, *Darwin's Metaphor: Nature's Place in Victorian Culture* (Cambridge: Cambridge University Press, 1985).

7 There are some notable exceptions to this as, for example: S. A. Barnett, 'The Expression of the Emotions', *A Century of Darwin*, ed. S. A. Barnett (Cambridge, Mass.: Harvard University Press, 1958), pp. 206–35; Janet Browne, 'Darwin and the Expression of the Emotions', *The Darwinian Heritage*, ed. David Kohn (Princeton: Princeton University Press, 1985); and 'Darwin and the Face of Madness', *The Anatomy of Madness: Essays in the History of Psychiatry*, ed. W. F. Bynum, Roy Porter and Michael Shepherd, 3 vols. (London: Tavistock Publications, 1985), I, pp. 151–65; Paul Ekman, ed., *Darwin and Facial Expression: A Century of Research in Review* (New York: Academic Press, 1973).

8 Evelleen Richards, 'Redrawing the Boundaries: Darwinian Science and Victorian Women Intellectuals', *Victorian Science in Context*, ed. Bernard Lightman (Chicago: University Of Chicago Press, 1997), p.119. On the same theme, see also Richards', 'Darwin and the Descent of Woman', *The Wider Domain of Evolutionary Thought*, ed. David Oldroyd and Ian Langham (Dordrecht: Reidel, 1983), pp. 57–111.

9 Darwin states quite firmly that he is referring to the third edition (1844) of Bell's work because the first edition (1806) 'is much inferior in merit, and does not include some of his more important views'.

10 Browne, 'Darwin and the Expression of the Emotions', p. 317.

11 Richards, *Evolutionary Theories of Mind and Behaviour*, p. 231.

12 The opening sentences of *Expression of Emotions* read: 'Many works have been written on expression, but a greater number on physiognomy – that is, on the recognition of a character through the study of the permanent form of the features. With this latter subject I am not here concerned' (p. 7).

13 Peter F. Stevens, 'Species: Historical Perspectives', *Keywords in Evolutionary Biology*, ed. Evelyn Fox Keller and Elisabeth Lloyd (Cambridge, Mass.: Harvard University Press, 1992), pp. 302–11, p. 304. See also Harriet Ritvo, *The Platypus and the Mermaid and Other Figments of the Classifying Imagination* (Cambridge, Mass.: Harvard University Press, 1997). Making the same point, Ritvo writes: 'the assumption that species names represented essentially real and unchanging entities did not greatly assist in [naturalists] identifying them on the wing or the hoof. Although, as Darwin later conceded, many species were "tolerably well-defined objects" . . . many species were demonstrably not' (p. 86).

14 Darwin explained in his autobiography: 'My first child was born on December 27th, 1839, and I at once commenced to make notes on the first dawn of the various expressions which he exhibited, for I felt convinced, even at this early period, that the most complex and fine shades of expression must all have had a gradual and natural origin. During the summer of the following year, 1840, I read Sir C. Bell's admirable work on expression, and this greatly increased the interest which I felt in the subject, though I could not at all agree with his belief

that various muscles had been specially created for the sake of expression. From this time forward I occasionally attended to the subject, both with respect to man and our domesticated animals. My book sold largely; 5267 copies having been disposed of on the day of publication.' *The Autobiography of Charles Darwin 1809–1882, with original omissions restored*, ed. Nora Barlow (London: Collins, 1958), p. 50.

15 'Appendix III. Darwin's Observations on his Children', *The Correspondence of Charles Darwin*, ed. Frederick Burkhardt and Sydney Smith *et al.*, 9 vols.(Cambridge: Cambridge University Press, 1985–92), IV, p. 411. It is interesting to compare this entry to one some six months later to gauge the extent of Darwin's experimental attitude toward his son: 'During last week, when cold water put in mouth & more especially some rhubarb, he made expression of disgust very plainly, accompanied (& made very comical,) by look of surprise & consideration in his eyes, not knowing what to make of it. – The expression is accompanied by form of mouth – allowing what is in the mouth to run out' (p. 417). See also Darwin's later paper which uses some of the same material – 'Biographical Sketch of an Infant', *Mind* (7 July 1877): 285–94.

16 Darwin, 'Appendix III. Darwin's Observations on his Children', p. 421.

17 Incidentally, many psychologists now claim that infants acquire the capacity to be conscious – that is, consciousness – between approximately 24 and 30 months of age.

18 For a helpful description of the cultural influences on and intellectual debts of Darwin's thought, see Edward Manier, *The Young Darwin and his Cultural Circle: A Study of Influences which Helped Shape the Language and Logic of the Theory of Natural Selection* (Dordrecht: Reidel Publishing Co., 1978).

19 See 'Notebook M', *Charles Darwin's Notebooks, 1836–44. Geology, Transmutation of Species, Metaphysical Enquiries*, ed. Paul H. Barrett, Peter J. Gautrey, Sandra Herbert, David Kohn and Sydney Smith (Cambridge: British Museum (Natural History) and Cambridge University Press, 1987), pp. 556–8. References to notebook 'M' and 'N' will be to this invaluable edition. For an excellent analysis of the development of Darwin's thought in these notebooks, see Howard E. Gruber and P. H. Barrett, eds., *Darwin on Man: A Psychological Study of Scientific Creativity; together with Darwin's Early and Unpublished Notebooks* (New York: E. P. Dutton, 1974).

20 There is another (equally important) edition of the notebooks, specifically devoted to these two books (M and N). See Paul Barrett, ed., *Metaphysics, Materialism and the Evolution of Mind: Early Writings of Charles Darwin* (Chicago: University of Chicago Press, 1980).

21 See Ernst Mayr, *The Growth of Biological Thought: Diversity, Evolution, and Inheritance* (Cambridge, Mass.: Harvard University Press, 1982).

22 Darwin, 'Notebook M', p. 556. Some months later, Darwin recalled with interest that his sister, Marianne, 'says she has constantly observed that very young children. express the greatest surprise at emotions in

her countenance – before they have learnt by experience, that movements of face are more expressive than movements of fingers'. See
'Notebook N', *Charles Darwin's Notebooks, 1836–44*, p. 573.

23 See Gruber and Barrett, *Darwin on Man*, p. 326.

24 Richards, *Evolutionary Theories of Mind and Behaviour*, p. 231.

25 Darwin, *The Expression of the Emotions*, p. 18.

26 Ibid., pp. 15–16.

27 See Edwin Clarke and L. S. Jacyna, *Nineteenth-Century Origins of Neuroscientific Concepts* (Berkeley: University of California Press, 1987),
pp. 157–211. Clarke and Jacyna provide a detailed description of the
rise of experimentation on nerve function, including Galvani's researches, to which I cannot hope to do justice in a few sentences. They
explain the background to the galvanisation of nerve impulses and the
related (and extremely fractious) debate about the connection of animal
electricity and magnetism in the hands of Anton Mesmer.

28 See William Montgomery, 'Charles Darwin's Thought on Expressive
Mechanisms in Evolution', *The Development of Expressive Behaviour: Biology–
Environment Interactions* (New York: Academic Press, 1985), pp. 27–50.

29 It is worth noting, incidentally, that in this Darwin arrived at something
close to the subsequent James Lange theory of emotions, as Howard
Gruber has explained: 'The external event to which the person or
animal responds normally evokes from him some overt action; this act
in turn has direct consequences for the physical state of the organism;
these become the concomitants of emotion even when the overt action
is omitted.' See Gruber and Barrett, *Darwin on Man*, p. 310.

30 Herbert Spencer, *Essays, Scientific, Political and Speculative* (London, 1863),
p. 111; cited in *The Expression of the Emotions*, p. 71.

31 Darwin intended these photographs to show how grief induced 'the
raising of the inner ends of the eyebrows, and the drawing down of the
corners of the mouth' (p. 177) due to the movement of facial muscles
which were linked to the emotional experience of grief. He explained:
'The grief-muscles are not very often brought into play; and as the
action is often momentary, it easily escapes observation. Although the
expression, when observed, is universally and instantly recognised as
that of grief or anxiety, yet no one person out of a thousand who has
never studied the subject, is able to say precisely what change passes
over the sufferer's face' (p. 183).

32 Responding to this negative review of *The Expression of the Emotions*
written by Bain, Darwin accepted the weakness of this principle. He
said: 'your criticisms are all written in a quite fair spirit, and indeed no
one who knows you or your work would expect anything else. What you
say about the vagueness of what I have called the direct action of the
nervous system is perfectly just. I felt it so at the time, and even more of
late'. Cited by Paul Ekman, ed., *The Expression of the Emotions* (London:
HarperCollins, 1998), p. 87.

33 Darwin, *The Expression of the Emotions*, p. 43.

34 See Darwin's notes on breeding and its relation to expression, and Robert B. Freeman and Paul J. Gautrey, 'Darwin's *Questions about the Breeding of Animals*, with a Note on *Queries about Expression*', *Journal of the Society for the Bibliography of Natural History* 5 (1969): 220–5; and 'Charles Darwin's *Queries about Expression*', *Bulletin of the British Museum (Natural History)* 4 (1972): pp. 207–19.

35 See Gruber, *Darwin on Man*, 266–75.

36 Richards, *Evolutionary Theories of Mind and Behaviour*, pp. 95–6.

37 Cited in Gruber, *Darwin on Man*, p. 274.

38 Ibid., p. 280.

39 In this, Richards' view of Darwin's early theories of species change differs considerably from the widely accepted account of David Kohn. See Kohn, 'Theories to Work By: Rejected Theories, Reproduction, and Darwin's Path to Natural Selection', *Studies in the History of Biology* 4 (1980): 67–170. Richards details the differences between Kohn and himself in an appendix to the second chapter on 'Behaviour and mind in evolution'; see pp. 124–6.

40 For an overview of Lamarck's ideas, see F. W. Burkhardt, *The Spirit of System: Lamarck and Evolutionary Biology* (Cambridge, Mass.: Harvard University Press, 1977) and Ludmilla J. Jordanova, *Lamarck* (Oxford: Oxford University Press, 1984).

41 Stephen Jay Gould, *Ontogeny and Phylogeny* (Cambridge, Mass.: The Belknap Press of Harvard University, 1977), p. 96.

42 Darwin's copy was as follows: Johannes Müller, *Elements of Physiology*, trans. W. Baly, 2 vols. (London: Taylor and Watson, 1839–42). He seems to have read the first volume in 1840, and found confirmation of his speculations on instinct in 'Notebook M'. For more detail on Müller's study, see Clarke and Jacyna, *Nineteenth-Century Origins*, pp. 124–6, 150–1, and 205–6.

43 The background to this process has been investigated by Roger Smith in a fascinating doctoral dissertation, 'Physiological Psychology and the Philosophy of Nature in Mid-Nineteenth-Century Britain' (University of Cambridge, 1971); see also Smith's book on *Inhibition: History and Meaning in the Sciences of Mind and Brain* (Berkeley: University of California Press, 1992), pp. 113–78. The following summary is indebted to the excellent work of Smith, Clarke and Jacyna on the subject.

44 See Darwin, *The Expression of the Emotions*, p. 47 on Carpenter in relation to reflex action, and p. 335 on Laycock in respect of the involuntary nature of blushing.

45 Carpenter, 'On the Voluntary and Instinctive Actions in Living Beings', *Edinburgh Medical and Surgical Journal* (1837): 24–5. The list of his works is extensive, but the most popular was *Principles of General and Comparative Physiology, intended as an introduction to the study of human physiology, and as a guide to the philosophical pursuit of natural history*

(London: John Churchill, 1839) – of which five editions had been produced by 1854.

46 Cited by Alison Winter, 'The Construction of Orthodoxies and Hetero-doxies in the Early Victorian Life Sciences', *Science in Context*, ed. Bernard Lightman (Chicago: University of Chicago Press, 1998), pp. 24–50, p. 39.

47 Carpenter, 'Voluntary and Instinctive Actions', pp. 26–7.

48 Clarke and Jacyna, *Nineteenth-Century Origins*, p. 140.

49 William Carpenter, '[Review of] *The Brain and its Physiology* by Daniel Noble', *British and Foreign Medical Review* 22 (1846): 513.

50 Thomas Laycock, *On the Reflex Function of the Brain* (London, 1844), p. 8.

51 Cited by Jacyna, 'The Physiology of Mind', p. 112.

52 *The Expression of the Emotions*, p. 45. He had earlier made a similar point in *The Origin of Species*: 'domestic instincts have been acquired and natural instincts have been lost partly by habit, and partly by man selecting and accumulating during successive generations, peculiar mental habits and actions, which at first appeared from what we must in our ignorance call an accident' (p. 176).

53 *The Expression of the Emotions*, p. 310.

54 Darwin explained the moral sense as follows: 'For each individual would have an inward sense of possessing certain stronger or more enduring instincts, and others less strong or enduring; so that there would often be a struggle which impulse should be followed; and satisfaction or dissatisfaction would be felt, as past impressions were compared during their incessant passage through the mind. In this case, an inward monitor would tell the animal that it would have been better to have followed the one impulse rather than the other: the one would have been right and the other wrong' (*The Descent of Man*, pp. 73–4).

55 Richards, *Evolutionary Theories of Mind and Behaviour*, p. 210.

56 Darwin, *The Descent of Man*, p. 87.

57 Darwin, 'Notebook N', pp. 567–8.

58 Charles Darwin, *Journal of Researches into the Geology and Natural History of the Various Countries Visited during the Voyage of H. M. S. Beagle Round the World under the Command of Capt. FitzRoy R. N. from 1832 to 1836* (London: Henry Colburn, 1839), p. 228.

59 Ibid., p. 231.

60 Darwin, 'Notebook N', p. 577.

61 Thomas Burgess, 'Preface', *The Physiology or Mechanism of Blushing; illustrative of the influence of mental emotion on the capillary circulation; with a general view of the sympathies, and the organic relations of those structures with which they seem to be connected* (London: John Churchill, 1839).

62 Burgess, *The Physiology*, p. 1.

63 Ibid., p. 24.

64 Darwin, *The Descent of Man*, p. 394.

65 *The Origin of Species*, p. 427.

6 THE PROMISE OF A NEW PSYCHOLOGY

1 Charles Darwin, *The Origin of Species*, ed. Gillian Beer (Oxford: World's Classics, 1998), p. 392.
2 Studies of Galton have tended to emphasise his eugenicist ideas rather than his use of physiognomic teachings; see, for example, Karl Pearson, *The Life, Letters, and Labours of Francis Galton*, 2 vols. (Cambridge: Cambridge University Press, 1924); D. W. Forrest, *Francis Galton: The Life and Work of a Victorian Genius* (London: Routledge, 1974); and Ruth Schwartz, *Sir Francis Galton and the Study of Heredity in the Nineteenth Century* (New York: Garland Publishing, 1984). The notable exceptions to this are: Daniel J. Kevles, *In the Name of Eugenics: Genetics and the Uses of Human Heredity* (Harmondsworth: Penguin, 1986) and Daniel Pick, *Faces of Degeneration: A European Disorder, c.1848–c.1918* (Cambridge: Cambridge University Press, 1989).
3 Francis Galton, *Inquiries into Human Faculty and its Development* (London: Macmillan and Co., 1883), pp. 24–5. In a lengthy note, Galton explained that the term 'eugenics' was useful in expressing 'the science of improving stock, which is by no means confined to questions of judicious mating, but which, especially in the case of man, takes cognisance of all influences that tend in however remote a degree to give to the more suitable races or strains of blood a better chance of prevailing over the less suitable than they otherwise would have had' (p. 25).
4 Ibid., pp. 1–2.
5 For the wider context to the relationship of Darwin to Lamarck, see Toby Appel, *The Cuvier–Geoffroy Debate: French Biology in the Decades before Darwin* (New York: Oxford University Press, 1987); F. W. Burkhardt, *The Spirit of the System: Lamarck and Evolutionary Biology* (Cambridge, Mass.: Harvard University Press, 1977); M. J. S. Hodge, 'Lamarck's Science of Living Bodies', *British Journal for the History of Science* 5 (1971): 323–52; Ludmilla J. Jordanova, *Lamarck* (Oxford: Oxford University Press, 1984); Ernst Mayr, ed., *Evolution and the Diversity of Life* (Cambridge, Mass.: Harvard University Press, 1976).
6 Galton published widely throughout his life on a number of related topics. See, in particular, *Hereditary Genius: An Inquiry into its Laws and Consequences* (London, 1869), 2nd edn (London: Macmillan & Co., 1892); 'Hereditary Talent and Character', *Macmillan's Magazine* 12 (1865): 157–66 and 318–27; 'Hereditary Improvement', *Fraser's Magazine* NS 7 (1873): 116–30; *English Men of Science: Their Nature and Nurture* (London, 1874); *Natural Inheritance* (London: Macmillan & Co, 1889); *Memories of My Life*, 3rd edn (London: Methuen, 1909); *Essays in Eugenics* (London: Eugenics Education Society, 1909).
7 On this subject, see John R. Durant, 'The Meaning of Evolution: Post-Darwinian Debates on the Significance for Man of the Theory of Evolution, 1858–1908', PhD diss. (University of Cambridge, 1977) and

Lyndsay Andrew Farrall, 'The Origins and Growth of the English Eugenics Movement 1865–1925', PhD diss. (University of Indiana, 1970).

8 Galton, 'Hereditary Talent and Character', p. 161.

9 Herbert Spencer, *The Principles of Psychology* (London: Longman, Brown, Green and Longmans, 1855), p. 539.

10 Ibid., p. 423.

11 Ibid., p. 486.

12 For a more detailed account of Spencer's work, see John W. Burrow, *Evolution and Society* (Cambridge: Cambridge University Press, 1966), chap. 6; John C. Greene, 'Biology and Social Theory in the Nineteenth Century: Auguste Comte and Herbert Spencer', *Critical Problems in the History of Science*, ed. M. Claggett (Madison: University of Wisconsin Press, 1959), pp. 419–46; Robert J. Richards, *Darwin and the Emergence of Evolutionary Theories of Mind and Behaviour* (Chicago: University of Chicago Press, 1987), chaps. 6–7; Robert M. Young, *Mind, Brain, and Adaptation in the Nineteenth Century* (Oxford: Oxford University Press, 1971), chap. 5.

13 See, in particular, Richards, *Evolutionary Theories of Mind and Behaviour* and Young, *Mind, Brain, and Adaptation*.

14 Charles Lyell, *The Geological Evidences of the Antiquity of Man* (London: John Murray, 1863), pp. 504–5. For more on the place of Lyell in these debates, see Peter J. Bowler, *Theories of Human Evolution: A Century of Debate, 1844–1944* (Baltimore: Johns Hopkins University Press, 1986); Charles C. Gillispie, *Genesis and Geology: A Study in the Relations of Scientific Thought, Natural Theology, and Social Opinion in Great Britain, 1790–1850* (Cambridge, Mass., 1951); N. A. Rupke, *The Great Chain of History: William Buckland and the English School of Geology, 1814–1849* (Oxford: Oxford University Press, 1983).

15 Galton, 'Hereditary Talent and Character', p. 165.

16 Ibid., p. 326.

17 See Robert M. Young, *Darwin's Metaphor: Nature's Place in Victorian Culture* (Cambridge: Cambridge University Press, 1985), chaps. 1 and 2; and Stephen Copley, 'The Natural Economy: A Note on Some Rhetorical Strategies in Political Economy – Adam Smith and Malthus', *1789: Reading Writing Revolution*, ed. Francis Barker (Colchester: University of Essex Press, 1982), pp. 160–9.

18 Galton, *Inquiries into Human Faculty*, p. 3.

19 Galton, 'Generic Images', *Nineteenth Century*, 6 (1879): pp. 162–63. For the companion piece to this essay, see 'On Generic Images: with autotype illustrations', *Proceedings of the Royal Institution*, 9 (1879): pp. 159–65, and also 'Analytical Photography', *Nature*, 18 (1890): 381–87.

20 Alan Sekula, 'The Body and the Archive', *October* 39 (1986): 3–64, p. 48.

21 Galton, *Inquiries into Human Faculty*, p. 337.

Bibliography

Allentuck, Maria. 'Fuseli and Lavater: Physiognomical Theory and the Enlightenment'. *Studies in Voltaire and the Eighteenth Century* 15 (1967): 89–112.

Angell, James Rowland. 'The Influence of Darwin on Psychology'. *Psychological Review.* 16 (1909): 152–69.

Appel, Toby. *The Cuvier-Geoffroy Debate: French Biology in the Decades Before Darwin.* New York: Oxford University Press, 1987.

Bain, Alexander. *Senses and the Intellect.* John W. Parker and Son, 1855.
The Emotions and the Will (1859). 2nd edn London: Longmans, 1865.

Ball, Patricia. *The Science of Aspects: The Changing Role of Fact in Coleridge, Ruskin, and Hopkins.* London: The Athlone Press, 1971.

Barasch, Mosche. *Giotto and the Language of Gesture.* Cambridge: Cambridge University Press, 1981.

Barish, Jonas. *The Antitheatrical Prejudice.* Berkeley: University of California Press, 1981.

Barlow, Nora, ed. *The Autobiography of Charles Darwin 1809–1882, with Original Omissions Restored.* London: Collins, 1958.

Barlow, Paul. 'Pre-Raphaelitism and Post-Raphaelitism: The Articulation of Fantasy and the Problem of Pictorial Space'. *Pre-Raphaelites Re-Viewed.* Ed. Marcia Pointon. Manchester: Manchester University Press, 1989. Pp. 66–82.

Barnes, Barry, and Steven Shapin, eds. *Natural Order: Historical Studies of Scientific Culture.* London: Sage, 1979.

Barnett, S. A. 'The Expression of the Emotions'. *A Century of Darwin.* Ed. S. A. Barnett. Cambridge, Mass.: Harvard University Press, 1958. Pp. 206–30.

La Barre, Weston. 'The Cultural Basis of Emotions and Gestures'. *Journal of Personality* 16 (1947–8): 49–68.

Barrett, P. H. *Metaphysics, Materialism and the Evolution of the Mind: Early Writings of Charles Darwin.* Chicago: University of Chicago Press, 1980.

Barrett, P. H., ed. *The Collected Papers of Charles Darwin.* 2 vols. Chicago: University of Chicago Press, 1977.

Barrett, P. H., Donald J. Weinshank, Paul Ruhlen, Stehan J. Ozminski and

Barbara N. Berghage, eds. *A Concordance to Darwin's "The Expression of the Emotions in Man and Animals"*. Ithaca: Cornell University Press, 1986.

Barrett, P. H., Peter J. Gautrey, Sandra Herbert, David Kohn, and Sydney Smith, eds. *Charles Darwin's Notebooks, 1836–1844. Geology, Transmutation of Species, Metaphysical Enquiries*. Cambridge: British Museum (Natural History) and Cambridge University Press, 1987.

Basalla, George, William Coleman, and Robert H. Kargon, eds. *Victorian Science: A Self-Portrait from the Presidential Addresses to the British Association for the Advancement of Science*. New York: Doubleday, 1970.

Beer, Gillian. *Darwin's Plots: Evolutionary Narrative in Darwin, George Eliot and Nineteenth-Century Fiction*. London: Routledge & Kegan Paul, 1983.

'Darwin's Reading and the Fictions of Development'. *The Darwinian Heritage*. Ed. David Kohn. Princeton: Princeton University Press, 1985. Pp. 543–88.

'Four Bodies on the *Beagle*: Touch, Sight and Writing in a Darwin Letter'. *Textuality & Sexuality: Reading Theories and Practices*. Ed. Judith Still and Michael Worton. Manchester: Manchester University Press, 1993. Pp. 116–32.

Bell, Charles. *Essays on the Anatomy of Expression in Painting*. London: George Bell & Sons, 1806.

A System of Dissections, Explaining the Anatomy of the Human Body, the Manner of Displaying the Parts, and their Varieties in Disease. 2nd edn London: Longman, Hurst, Rees & Orme, 1810.

'Idea of a New Anatomy of the Brain'. *Philosophical Transactions of the Royal Society of London* (1811). *The Way In and the Way Out: François Magendie, Charles Bell, and the Roots of the Spinal Nerves*. Ed. Paul F. Cranefield. New York: Futura, 1974.

'On the Nerves; giving an account of some experiments on their structure and functions, which lead to a new arrangement of the system'. *Philosophical Transactions of the Royal Society of London* 111 (1821): 398–424.

'On the Nerves of the Face; being a second paper on that subject'. *Philosophical Transactions of the Royal Society of London*. London, 1829. *Selections from the Writings of Sir Charles Bell. Medical Classics* I (1936): 154–69.

The Hand: Its Mechanisms and Vital Endowments as Evincing Design. London: William Pickering, 1833.

Animal Mechanics, or, Proofs of Design in the Animal Frame, Library of Useful Knowledge, Natural Philosophy IV. London: Baldwin & Craddock, 1838.

Familiar Treatises on the Five Senses; being an account of the conformation and functions of the eye, ear, nose, tongue and skin, illustrated by twenty colour plates. London: Henry Washbourne, 1841.

The Anatomy and Philosophy of Expression as connected with the Fine Arts. 3rd ed. London: George Bell & Sons, 1844.

Letters of Sir Charles Bell. Ed. G. J. Bell. London: John Murray, 1870.

Bignamini, Ilaria and Martin Postle, eds. *The Artist's Model: Its Role in British Art from Lely to Etty.* Nottingham: Nottingham University Press, 1991.

Bowler, Peter J. *Evolution: The History of an Idea.* Berkeley: University of California Press, 1984.

    Theories of Human Evolution: A Century of Debate, 1844–1944. Baltimore: Johns Hopkins University Press, 1986.

Bowness, Alan, ed. *The Pre-Raphaelites.* London: Tate Gallery and Penguin Books, 1984.

Brand, Dana Aaron. *The Spectator and the City: Fantasies of Urban Legibility in Nineteenth-Century England and America.* Ann Arbor: University Microfilms International, 1986.

Brantlinger, Patrick. 'What is Sensational about the Sensation Novel?' *Nineteenth Century Fiction* 37 (1982): 1–28.

Brantlinger, Patrick, ed. *Energy and Entropy: Science and Culture in Victorian Britain.* Bloomington: Indiana University Press, 1989.

Bremmer, Jan and Herman Roodenburg, eds. *A Cultural History of Gesture: From Antiquity to the Present Day.* Cambridge: Polity Press, 1991.

Broadbent, R. J. *A History of Pantomime.* New York: Benjamin Bloom, 1964.

Brown, Ford Madox. *The Exhibition of Work and Other Paintings.* London: McCorquodale & Co., 1865.

    The Diary of Ford Madox Brown. Ed. V. Surtees. New Haven: Yale University Press, 1981.

Brown, Richard. *An Essay on the Truth of Physiognomy and its Application to Medicine.* Philadelphia, 1807.

Browne, Janet. 'Darwin and the Face of Madness'. *The Anatomy of Madness.* Ed. W. F. Bynum, Roy Porter and Michael Shepherd. 3 vols. London: Tavistock Publications, 1985. I, pp. 151–65.

    'Darwin and the Expression of the Emotions'. *The Darwinian Heritage.* Ed. David Kohn. Princeton: Princeton University Press, 1985.

Bruce, Vicki. *Recognising Faces.* Hove: Lawrence Erlbaum Associates, 1988.

Bryson, Norman. *Vision and Painting: The Logic of the Gaze.* Houndmills: Macmillan Press, 1983.

Burgess, S.W. *Historical Illustrations of the Origin and Progress of the Passions, and their Influence on the Conduct of Mankind, with some subordinate sketches of human nature and human life.* London: Longman, 1825.

Burgess, T. H. *The Physiology or Mechanism of Blushing; Illustrative of the Influence of Mental Emotion on the Capillary Circulation; with a General View of the Sympathies, and the Organic Relations of those Structures with which they seem to be connected.* London: John Churchill, 1839.

Burkhardt, Frederick W. *The Spirit of System: Lamarck and Evolutionary Biology.* Cambridge, Mass.: Harvard University Press, 1977.

    'Darwin on Animal Behaviour and Evolution'. *The Darwinian Heritage.* Ed. David Kohn. Princeton: Princeton University Press, 1985.

Burkhardt, Frederick W. and Sydney Smith, eds. *The Correspondence of Charles Darwin.* 9 vols. Cambridge: Cambridge University Press, 1985–1992.

Burnett, J. *Practical Hints on Portrait Painting, illustrated by examples from the works of Van Dyck and other masters*. London, 1850.

Burrow, John W. *Evolution and Society: A Study in Victorian Social Theory.* Cambridge: Cambridge University Press, 1966.

Butler, Joseph. *The Analogy of Religion Natural and Revealed to the Constitution and Course of Nature*. London: Robert Carter & Bros., 1736.

Bynum, William F. and Roy Porter, eds. *Medicine and the Five Senses*. Cambridge: Cambridge University Press, 1993.

Bynum, William F., Roy Porter and Michael Shepherd, eds. *The Anatomy of Madness: Essays in the History of Psychiatry*. 3 vols. London: Routledge, 1985–88.

Calderwood, H. 'The Expression of Emotions on the Human Countenance'. *International Review* 6 (1878): 195–204.

Camper, Pieter. *The Works of the late Professor Camper on the Connexion between the Science of Anatomy and the Arts of Drawing, Painting, Statuary, etc., etc*. Ed. T. Cogan. 4th edn London: J. Hearne, 1821.

Cannon, Susan F. *Science in Culture: The Early Victorian Period*. New York: Dawson and Science History Publications, 1978.

Cannon, W. B. 'The James-Lange Theory of Emotion: A Critical Examination and an Alternative Theory'. *American Journal of Psychology* 39 (1927): 106–24.

Carlisle, Anthony. '[The Relation between Anatomy and the Artist's Training]'. *The Artist*, 1807.

Carpenter, W. B. 'On the Voluntary and Instinctive Actions', *Edinburgh Medical and Surgical Journal* (1837): 24–5.

Principles of General and Comparative Physiology, intended as an introduction to the study of human physiology; and as a guide to the philosophical pursuit of natural history. London: John Churchill, 1839.

Principles of Human Physiology (1842). 4th edn London: John Churchill, 1853.

Clarke, Rev. W. T. *The Phrenological Miscellany; or, The Annuals of Phrenology and Physiognomy, from 1865 to 1873*. New York: Fowler and Wells, 1887.

Codell, Julie F. 'Expression over Beauty: Facial Expression, Body Language, and Circumstantiality in the Paintings of the Pre-Raphaelite Brotherhood'. *Victorian Studies* 29 (1986): 255–90.

'The Dilemma of the Artist in Millais' *Lorenzo and Isabella*: Phrenology, the Gaze and the Social Discourse'. *Art History* 14 (1991): 51–66.

Cohen, I. Bernard. 'Three Notes on the Reception of Darwin's Ideas on Natural Selection (Henry Baker Tristram, Alfred Newton, Samuel Wilberforce)'. *The Darwinian Heritage*. Ed. David Kohn. Princeton: Princeton University Press, 1985. Pp. 589–607.

Coleman, W. *Georges Cuvier, Zoologist. A Study in the History of Evolution Theory.* Cambridge, Mass.: Harvard University Press, 1964.

Biology in the Nineteenth Century: Problems of Form, Function, and Transformation. New York: John Wiley, 1972.

Collins, Wilkie. *Basil. A Story of Modern Life* (1852). New York: Dover Publications, 1980.

The Woman in White (1860). Ed. Harvey Peter Sucksmith. Oxford: Oxford University Press, 1973.

No Name (1862). Ed. Virginia Blain. Oxford: Oxford University Press, 1986.

My Miscellanies. In Two Volumes. London: Sampson, Low, Son & Co., 1863.

The Moonstone (1868). Ed. Anthea Trodd. Oxford: Oxford University Press, 1982.

'Books Necessary for a Liberal Education'. *Pall Mall Gazette* (1886): 2.

'How I Write my Books: Related in a Letter to a Friend'. *The Globe* (1887): 6.

Cooke, T. *A Practical and Familiar View of the Science of Physiognomy.* London, 1820.

Cooper, Robyn. 'The Relationship between the Pre-Raphaelite Brotherhood and Painters before Raphael in English Criticism of the late 1840s and 1850s'. *Victorian Studies* 24 (1981): 405–38.

'Definition and Control: Alexander Walker's Trilogy on Woman'. *History of Sexuality* 2 (1992): 341–64.

'Victorian Discourses on Women and Beauty: The Alexander Walker Texts'. *Gender and History* 5 (1993): 34–55.

Cooter, Roger. *The Cultural Meaning of Popular Science: Phrenology and the Organisation of Consent in Nineteenth-Century Britain.* Cambridge: Cambridge University Press, 1985.

Copley, Stephen. 'The Natural Economy: A Note on Some Rhetorical Strategies in Political Economy – Adam Smith and Malthus'. *1789: Reading Writing Revolution.* Ed. Francis Barker. Proceedings of Essex Conference on the Sociology of Literature, July 1981. Colchester: University of Essex Press, 1982. Pp. 160–9.

Cosslett, Tess. *The 'Scientific Movement' and Victorian Literature.* Brighton: Harvester Press, 1982.

Cosslett, Tess, ed. *Science and Religion in the Nineteenth Century.* Cambridge: Cambridge University Press, 1984.

Cowan, Ruth Schwartz. *Sir Francis Galton and the Study of Heredity in the Nineteenth Century.* New York: Garland Publishing, 1985.

Cowling, Mary. *The Artist as Anthropologist: The Representation of Type and Character in Victorian Art.* Cambridge: Cambridge University Press, 1989.

Cowper, William. 'The History of Physiognomy'. *Memoirs of the Manchester Literary Society.* Vol. II. Manchester, *c.*1791.

Cranefield, Paul F. *The Way In and the Way Out: François Magendie, Charles Bell, and the Roots of the Spinal Nerves.* New York: Futura, 1974.

Cross, John. *An Attempt to Establish Physiognomy upon Scientific Principles.* London, 1817.

Cummings, Frederick. 'B. R. Haydon and His School'. *Journal of Warbourg and Courtauld Institutes.* 26 (1963): 367–80.

'Charles Bell's *Anatomy of Expression*'. *Art Bulletin* 46 (1964): 191–203.

Cvetkovich, Ann. 'Ghostlier Determinations: The Economy of Sensation and *The Woman in White*'. *Novel* 23 (1989): 24–43.

Mixed Feelings: Feminism, Mass Culture, and Victorian Sensationalism. New Brunswick: Rutgers University Press, 1992.

Dale, Peter Allen. *In Pursuit of a Scientific Culture: Science, Art, and Society in the Victorian Age.* Madison: University of Wisconsin Press, 1989.

[Dallas, E. S.]. 'On Physiognomy'. *Cornhill Magazine* 4 (1861): 472–81.

'The First Principle of Physiognomy'. *Cornhill Magazine* 4 (1861): 569–81.

Darwin, Charles R. *Journal of Researches into the Geology and Natural History of the Various Countries Visited during the Voyage of H. M. S. Beagle Round the World under the Command of Capt Fitz Roy, R. N. from 1832 to 1836.* London: Henry Colburn, 1839.

On the Origin of Species by Means of Natural Selection, or the Preservation of Favoured Races in the Struggle for Life. London: John Murray, 1859.

The Descent of Man and Selection in Relation to Sex. (1871). Princeton: Princeton University Press, 1981.

The Expression of the Emotions in Man and Animals. (1872). Chicago: University of Chicago Press, 1965.

'Biographical Sketch of an Infant'. *Mind* (7 July 1877): 285–94.

On the Origin of Species, by Charles Darwin; A Variorum Text. Ed. M. Peckham. Philadelphia: University of Pennsylvania Press, 1959.

Charles Darwin's Natural Selection being the Second Part of his Big Species Book written from 1856 to 1858. Ed. R. Stauffer. 2 vols. London: Cambridge University Press, 1975.

The Beagle Record: Selections from the Original Pictorial Records and Written Accounts of the Voyage of H.M.S. Beagle. Ed. Richard Darwin Keynes. Cambridge: Cambridge University Press, 1979.

The Expression of the Emotions in Man and Animals. Definitive Edition. Ed. Paul Ekman. London: HarperCollins, 1998.

Darwin, Charles R. and Alfred R. Wallace. *Evolution by Natural Selection.* Cambridge: Cambridge University Press, 1958.

Darwin, Erasmus. *The Temple of Nature; or, The Origin of Society, a poem, with philosophical notes.* London, 1803.

Darwin, Francis, ed. *The Life and Letters of Charles Darwin, including an Autobiographical Chapter.* 3 vols. London: John Murray, 1887.

The Autobiography of Charles Darwin and Selected Letters. New York: Dover Publications, 1958.

Darwin, Francis and A. C. Seward, ed. *More Letters of Charles Darwin; A record of his work in a series of hitherto unpublished letters.* 2 vols. London: John Murray, 1903.

Daston, Lorraine. 'British Responses to Psycho-Physiology, 1860–1900'. *Isis* 69 (1978): 192–208.

'The Theory of Will versus Science of Mind'. *The Problematic Science: Psychology in Nineteenth-Century Thought.* Ed. W. R. Woodward and Mitchell G. Ash. New York: Praeger, 1982. Pp. 88–115.

Desmond, Adrian. 'Lamarckism and Democracy: Corporations, Corruption and Comparative Anatomy in the 1830s'. *History, Humanity and Evolution: Essays for John C. Greene.* Ed. James R. Moore. Cambridge: Cambridge University Press, 1989. Pp. 99–130.

    The Politics of Evolution: Morphology, Medicine, and Reform in Radical London. Chicago: University of Chicago Press, 1989.

Desmond, Adrian and James R. Moore. *Darwin.* Harmondsworth: Penguin Books, 1993.

Dewey, John. 'The Theory of Emotion (1)'. *Psychological Review* 1 (1894): 553–69.

    'The Theory of Emotion (2)'. *Psychological Review* 2 (1895): 13–32.

Dowling, Linda. *The Vulgarization of Art: The Victorians and Aesthetic Democracy.* Charlottesville: University of Virginia Press, 1996.

Durant, John R. 'The Meaning of Evolution: Post-Darwinian Debates on the Significance for Man of the Theory of Evolution, 1858–1908'. PhD diss., University of Cambridge, 1977.

[Eastlake, Elizabeth S.]. 'Physiognomy'. *Quarterly Review* 90 (December 1851): 62–91.

Ekman, Paul. 'Facial Signs: Facts, Fantasies, and Possibilities'. *Sight, Sound and Sense.* Ed. T. Sebeok. Bloomington: Indiana University Press, 1978.

Ekman, Paul, ed. *Darwin and Facial Expression: A Century of Research in Review.* New York: Academic Press, 1973.

Ekman, Paul, W. Friesen, and P. Ellsworth. *Emotion in the Human Face.* 2nd edn Cambridge: Cambridge University Press, 1982.

Ellegard, A. *Darwin and the General Reader: The Reception of Darwin's Theory of Evolution in the British Periodical Press, 1859–1872.* Gothenburg Studies in English, no. 8. Stockholm: Almqvist and Wiksel, 1958.

Ellis, Steve. 'Rossetti and the Cult of the *Vita Nuova*'. *Dante and English Poetry: Shelley to T.S. Eliot.* Cambridge: Cambridge University Press, 1983. Pp. 102–139.

Errington, Lindsay. *Social and Religious Themes in English Art, 1840–1860.* New York: Garland Publishing, 1984.

Evans, Elizabeth C. 'Physiognomics in the Ancient World'. *Transactions of the American Philosophical Society* 59 (1969): 3–101.

Faas, Ekbert. *Retreat into the Mind: Victorian Poetry and the Rise of Psychiatry.* Princeton: Princeton University Press, 1988.

Fahnestock, Jeanne. 'The Heroine of Irregular Features: Physiognomy and Conventions of Heroine Description'. *Victorian Studies* 24 (1981): 325–50.

Figlio, Karl. 'Theories of Perception and the Physiology of Mind in the Late Eighteenth Century'. *History of Science* 12 (1975): 177–212.

'The Metaphor of Organization: An Historiographical Perspective on the Bio-Medical Sciences of the Early Nineteenth Century'. *History of Science* 14 (1976): 17–53.

Ford, Ford Madox (Heuffer). *Ford Madox Brown, A Record of His Life and Work.* London: Longman's Green, 1896.

Ford, G. H. *Keats and the Victorians: A Study of his Influence and Rise to Fame, 1821–1895.* London: Yale University Press, 1944.

Forrest, D. W. *Francis Galton: The Life and Work of a Victorian Genius.* London: Routledge, 1974.

Fox, Christopher. 'Defining Eighteenth-Century Pyschology'. *Psychology and Literature in the Eighteenth Century.* Ed. Christopher Fox. (New York: AMS Press, 1987).

Fraser, Hilary. *Beauty and Belief: Aesthetics and Religion in Victorian Literature.* Cambridge: Cambridge University Press, 1986.

Fredeman, W. E. *Pre-Raphaelitism: A Bibliocritical Study.* Cambridge, Mass.: Harvard University Press, 1965.

Freeman, Robert B. and Paul J. Gautrey. 'Darwin's *Questions about the breeding of animals,* with a note on *Queries about expression'. Journal of the Society for the Bibliography of Natural History* 5 (1969): 220–5.

'Charles Darwin's *Queries about expression'. Bulletin of the British Museum (Natural History)* 4 (1972): 207–19.

Fried, Michael. *Absorption and Theatricality: Painting and Beholder in the Age of Diderot.* Berkeley: University of California Press, 1980.

Galton, Francis. 'Hereditary Talent and Character'. *Macmillan's Magazine* 12 (1865): 318–27.

Heredity Genius: An Inquiry into its Laws and Consequences (1869). 2nd edn. London: Macmillan, 1892.

'Hereditary Improvement'. *Fraser's Magazine.* NS 7 (1873): 116–30.

English Men of Science: Their Nature and Nurture. London, 1874.

'Generic Images'. *Nineteenth Century* 6 (1879): 157–69.

'On Generic Images; with autotype illustrations'. *Proceedings of the Royal Institution* 9 (1879): 57–61.

Inquiries into Human Faculty and its Development. London: Macmillan, 1883.

Natural Inheritance. London: Macmillan, 1889.

'Analytical Photography'. *Nature* 18 (1890): 381–7.

Memories of My Life. 3rd edn London: Methuen, 1909.

Essays in Eugenics. London: Eugenics Education Society, 1909.

Gere, J. A. *Pre-Raphaelite Drawings in the British Museum.* London: British Museum Press, 1994.

Ghiselin, M. T. *The Triumph of the Darwinian Method.* Berkeley: University of California Press, 1969.

Gillispie, Charles C. *Genesis and Geology: A Study in the Relations of Scientific Thought, Natural Theology, and Social Opinion in Great Britain, 1790–1850.* Cambridge, Mass.: Harvard University Press, 1951.

Gilman, Sander L. 'Touch, Sexuality and Disease'. *Medicine and the Five*

Senses. Ed. William F. Bynum and Roy Porter. Cambridge: Cambridge University Press, 1993. Pp. 198–224.

Health and Illness: Images of Difference. (London: Reaktion Books, 1995).

Giustino, D. de. *Conquest of the Mind: Phrenology and Victorian Social Thought.* London: Croom Helm, 1975.

Gombrich, Ernst H. 'The Mask and the Face: The Perception of Physiognomic Likeness in Life and Art'. E. H. Gombrich, Julian Hochberg and Max Black, *Art, Perception and Reality.* Baltimore: The Johns Hopkins University Press, 1972.

Gordon-Taylor, Gordon. *Sir Charles Bell, His Life and Times.* Edinburgh: E. & S. Livingston, 1958.

Gould, Stephen Jay. *Ontogeny and Phylogeny.* Cambridge, Mass.: Harvard University Press, 1977.

Graham, John. 'Lavater's *Physiognomy* in England'. *Journal of the History of Ideas* 22 (1961): 561–72.

'Character Description and Meaning in the Romantic Novel'. *Studies in Romanticism* 5 (1966): 208–18.

Lavater's 'Essays on Physiognomy': A Study in the History of Ideas. Berne: Peter Lang, 1979.

Grave, S. *The Scottish Philosophy of Common Sense.* Oxford: Oxford University Press, 1960.

Gray, Richard T. 'The Transcendence of the Body in the Transparency of its En-Signment: Johann Kaspar Lavater's "Physiognomical Surface Hermeneutics" and the Ideological (Con)Text of Bourgeois Modernism'. *Lessing Yearbook* 23 (1991): 127–48.

Green, David. 'Veins of Resemblance: Photography and Eugenics'. *The Oxford Art Journal* 7 (1984): 3–16.

Greene, John C. 'Biology and Social Theory in the Nineteenth Century: Auguste Comte and Herbert Spencer'. *Critical Problems in the History of Science.* Ed. M. Claggett. Madison: University of Wisconsin Press, 1959. Pp. 419–46.

The Death of Adam: Evolution and its Impact on Western Thought. New York: Mentor, 1961.

Science, Ideology and World View. Berkeley: University of California Press, 1981.

Grilli, Stephanie. 'Pre-Raphaelite Portraiture, 1848–1854'. PhD diss., Yale University, 1980.

Gruber, H. E. and P. H. Barrett. *Darwin on Man: A Psychological Study of Scientific Creativity; together with Darwin's Early and Unpublished Notebooks.* New York: E. P. Dutton, 1974.

Hamilton, Walter. *The Aesthetic Movement in England.* New York: AMS Press, 1971.

Haskell, Francis and Nicholas Penny. *Taste and the Antique: The Lure of Classical Sculpture, 1500–1900.* New Haven: Yale University Press, 1981.

Haydon, Benjamin Robert. *Lectures on Painting and Design.* London, 1846.
  The Life of Benjamin Robert Haydon, Historical Painter, from his Autobiography and Journals. Ed. T. Taylor. 2nd edn. 3 vols. London: Longman, Brown, Green & Longmans, 1853.
  The Autobiography and Journals of Benjamin Robert Haydon. Ed. M. Elwin. London: Macdonald, 1950.
Heller, Tamar. *Dead Secrets: Wilkie Collins and the Female Gothic.* New Haven: Yale University Press, 1992.
Helsinger, Elisabeth. *Ruskin and the Art of the Beholder.* Cambridge, Mass.: Harvard University Press, 1982.
Hewison, Robert. *John Ruskin, The Argument of the Eye.* Princeton: Princeton University Press, 1976.
Hilgard, Ernest. L. 'Psychology after Darwin'. *Evolution After Darwin. Vol. II. The Evolution of Man.* Ed. Sol Tax. Chicago: University of Chicago Press, 1960. Pp. 269–87.
Hilton, Timothy. *The Pre-Raphaelites.* London: Thames and Hudson, 1991.
Hipple, Walter. *The Beautiful, the Sublime, and the Picturesque in Eighteenth-Century British Aesthetic Theory.* Carbondale: Southern Illinois University Press, 1957.
Hughes, Winifred. *The Maniac in the Cellar: Sensation Novels of the 1860s.* Princeton: Princeton University Press, 1980.
Hull, David. *Darwin and his Critics: The Reception of Darwin's Theory of Evolution by the Scientific Community.* Cambridge, Mass.: Harvard University Press, 1973.
  Science as Process: An Evolutionary Account of the Social and Conceptual Development of Science. Chicago: Chicago University Press, 1988.
Hunt, John Dixon. *The Pre-Raphaelite Imagination:1848–1900.* London: Routledge & Kegan Paul, 1968.
Hunt, Leigh. 'On Pantomime' and 'On Pantomime, continued from a late paper'. *The Examiner* (5 and 27 January 1817). In *Leigh Hunt's Dramatic Criticism 1808–1881.* Ed. L. H. and C. W. Houtchens. London: Oxford University Press, 1950.
Hunt, William Holman. *Pre-Raphaelitism and the Pre-Raphaelite Brotherhood.* 2 vols. London, 1905.
Huxley, Thomas H. 'Time and Life: Mr. Darwin's *Origin of Species*'. *Macmillan's Magazine* 1 (1860): 142–8.
Inkster, Ian and Jack Morrell, eds. *Metropolis and Province: Science in British Culture 1780–1850.* London: Hutchison, 1983.
Irvine, William. *Apes, Angels and Victorians: The Story of Darwin, Huxley, and Evolution.* New York: McGraw-Hill, 1955.
Jack, Ian. 'Physiognomy, Phrenology and Characterisation in the Novels of Charlotte Brontë'. *Brontë Society Transactions* 15 (1970) 377–91.
Jacyna, L. S. 'The Physiology of Mind, the Unity of Nature, and the Moral Order in Victorian Thought'. *British Journal of the History of Science* 14 (1981): 109–32.

'Immanence or Transcendence: Theories of Life and Organization in Britain'. *Isis* 74 (1983): 311–29.

'Principles of General Physiology: The Comparative Dimension to British Neuroscience in the 1830s and 1840s'. *Studies in the History of Biology* 7 (1984): 47–92.

'The Romantic Programme and the Reception of Cell Theory in Britain'. *Journal of the History of Biology* 17 (1984): 13–48.

Jardine, Nick, James Secord, and Emma Spary, eds. *Cultures of Natural History.* Cambridge: Cambridge University Press, 1996.

Jay, W. 'Charles Darwin: Photography and Everything Else'. *British Journal of Photography* 7 (1980): 1116–18.

Johnson, Lewis. 'Pre-Raphaelitism, Personification, Portraiture'. *Pre-Raphaelites Re-Viewed*. Ed. Marcia Pointon. Manchester: Manchester University Press, 1989. Pp. 140–58.

Jordanova, Ludmilla J. *Sexual Visions: Images of Gender in Science and Medicine between the Eighteenth and Twentieth Centuries.* Hemel Hempstead: Harvester Wheatsheaf, 1989.

'Nature's Powers: A Reading of Lamarck's Distinction between Creation and Production'. *History, Humanity & Evolution: Essays for John C. Greene.* Ed. James R. Moore. Cambridge: Cambridge University Press, 1989. Pp. 71–98.

'The Hand'. *Visual Anthropology Review* 8 (1992): 2–7.

'The Art and Science of Seeing in Medicine: Physiognomy 1780–1820'. *Medicine and the Five Senses*. Ed. William F. Bynum and Roy Porter. Cambridge: Cambridge University Press, 1993. Pp. 122–33.

'The Representation of the Body: Art and Medicine in the Work of Charles Bell'. *Towards a Modern Art World. Studies in British Art I.* Ed. Brian Allen. New Haven, Conn.: Yale University Press, 1995. Pp. 79–94.

Jordanova, Ludmilla J., ed. *Languages of Nature: Critical Essays on Science and Literature.* London: Free Association Books, 1986.

Jordanova, Ludmilla J. and Roy Porter, eds. *Images of the Earth: Essays in the History of the Environmental Sciences.* Chalfont St. Giles: British Society for the History of Science, 1979.

Keats, John. *Poetical Works.* Ed. H. W. Garrod. Oxford: Oxford University Press, 1956.

Kemp, Martin. *The Science of Art: Optical Themes in Western Art from Brunelleschi to Seurat.* New Haven: Yale University Press, 1990.

Kendrick, Walter M. 'The Sensationalism of *The Woman in White*'. *Nineteenth Century Fiction* 32 (1977): 18–35.

Klonk, Charlotte. *Science and the Perception of Nature: British Landscape Art in the late Eighteenth and early Nineteenth Century.* New Haven: Yale University Press, 1996.

Knight, David. *The Age of Science: The Scientific World-View in the Nineteenth Century.* Oxford: Basil Blackwell, 1986.

Knoepflmacher, U. C. and G. B. Tennyson, eds. *Nature and the Victorian Imagination*. Berkeley: University of California Press, 1977.

Kohn, David, ed. *The Darwinian Heritage*. Princeton: Princeton University Press, 1985.

Krasner, James. 'A Chaos of Delight: Perception and Illusion in Darwin's Scientific Writing'. *Representations* 31 (1990): 118–41.

Kuhn, Thomas S. *The Structure of Scientific Revolutions*. Chicago: University of Chicago Press, 1962.

Lamarck, Jean-Baptiste. *Zoological Philosophy* (1809). Trans. Hugh Elliot. New York: Hafner Publishing, 1960.

Landow, George. *Victorian Types, Victorian Shadows*. Boston: Routledge & Kegan Paul, 1980.

Lavater, Johann Caspar. *Essays on Physiognomy; for the promotion of the Knowledge and the love of mankind*. Trans. and ed. Thomas Holcroft. 3 vols. London: G. G. & J. Robinson, 1789–93.

Essays on Physiognomy: Designed to Promote the Knowledge and Love of Mankind. Trans. and ed. H. Hunter. 2nd edn. 3 vols. London: John Stockdale, 1810.

Le Brun, Charles. *The Conference of Monsieur Le Brun, Chief Painter to the French King, Chancellor and Director of the Academy of Painting and Sculpture upon Expression, General and Particular, translated from the French*. London, 1707. (Trans. of *Conférences . . . sur l'expression générale et particulière*. Paris, 1698.)

A Method to Learn to Design the Passions. Intro. Q. T. McKenzie. Los Angeles: William Andrews Clark Memorial Library, University of California, Augustan Reprint Society, 1980.

Levine, George. *Darwin and the Novelists: Patterns of Science in Victorian Fiction*. Cambridge, Mass.: Harvard University Press, 1988.

Lewes, G. H. 'Mr. Darwin's Hypotheses. Part I' and 'Mr. Darwin's Hypotheses. Part II'. *The Fortnightly Review* 3 (1868): 353–73 and 611–28.

Liggett, John. *The Human Face*. London: Constable & Co., 1974.

Litvak, Joseph. *Caught in the Act: Theatricality in the Nineteenth-Century English Novel*. Berkeley: University of California Press, 1992.

Loesberg, Jonathan. 'The Ideology of Narrative Form in Sensation Fiction'. *Representations* 13 (1986): 115–38.

Lyell, Charles. *The Geological Evidences of the Antiquity of Man, with Remarks on Theories of the Origin of Species by Variation*. London: John Murray, 1863.

Lynn, Eliza. 'Passing Faces'. *Household Words: A Weekly Journal, conducted by Charles Dickens*. 11 (1855): 241–64.

MacCormack, C. P. and M. Strathern, eds. *Nature, Culture and Gender*. Cambridge: Cambridge University Press, 1980.

McMaster, Juliet. *The Index of the Mind: Physiognomy in the Novel*. Lethbridge: University of Lethbridge Press, 1990.

MacVicar, Rev. John G. *On the Beautiful, the Picturesque, the Sublime*. London: Scott, Webster & Geary, 1837.

Manier, Edward. *The Young Darwin and His Cultural Circle. A Study of Influences which Helped Shape the Language and Logic of the Theory of Natural Selection*. Dordrecht: D. Reidel Publishing Company, 1978.

Mansel, H. D. 'Sensation Novels'. *Quarterly Review* 113 (1863): 481–514.

Marsh, Jan. *Pre-Raphaelite Women: Images of Femininity in Pre-Raphaelite Art*. London: Weidenfeld & Nicolson, 1987.

Masson, David. 'Pre-Raphaelitism in Art and Literature'. *British Quarterly Review* 16 (1852): 197–220.

Mayr, Ernst. *The Growth of Biological Thought: Diversity, Evolution, and Inheritance*. Cambridge, Mass.: Harvard University Press, 1982.

Mayr, Ernst, ed. *Evolution and the Diversity of Life*. Cambridge, Mass.: Harvard University Press, 1976.

Meisel, Martin. ' "Half Sick of Shadows": The Aesthetic Dialogue in Pre-Raphaelite Painting'. *Nature and the Victorian Imagination*. Ed. U. C. Knoepflmacher and G. B. Tennyson. Berkeley: University of California Press, 1977.

Realizations: Narrative, Pictorial, Theatrical Arts in Nineteenth-Century England. Princeton: Princeton University Press, 1983.

Miller, D. A. 'From *Roman Policier* to *Roman-Police*: Wilkie Collins' *The Moonstone*'. *Novel* 13 (1980): 153–70.

'*Cage aux Folles:* Sensation and Gender in Wilkie Collins' *The Woman in White*'. *Representations* 14 (1986): 107–36.

Montagu, Jennifer. *The Expression of the Passions: The Origin and Influence of Charles Le Brun's 'Conference sur l'expression générale et particulière'*. New Haven, Conn.: Yale University Press, 1994.

Moore, James. *The Post-Darwinian Controversies: A Study of the Protestant Struggle to Come to Terms with Darwin in Great Britain and America, 1870–1900*. Cambridge: Cambridge University Press, 1979.

Moore, James, ed. *History, Humanity, and Evolution: Essays for John C. Greene*. Cambridge: Cambridge University Press, 1989.

Morrell, Jack and Arnold Thackray. *Gentlemen of Science: Early Years of the British Association for the Advancement of Science*. Oxford: Oxford University Press, 1981.

Munby, A. N. L. 'The Bibliophile: B. R. Haydon's Anatomy Book'. *Apollo* 26 (1937): 345–7.

Nead, Lynda. *Myths of Sexuality: Representations of Women in Victorian Britain*. Oxford: Oxford University Press, 1988.

Nord, Deborah Epstein. 'The City as Theatre: From Georgian to Early Victorian London'. *Victorian Studies* 31 (1988): 159–88.

Oldroyd, D. R. *Darwinian Impacts: An Introduction to the Darwinian Revolution*. Milton Keynes: The Open University Press, 1980.

[Oliphant, Margaret]. 'Modern Novelists – Great and Small'. *Blackwood's Magazine* 78 (1855): 554–68.

'Sensation Novels'. *Blackwood's Magazine* 91 (1862): 564–84.
'[Review of] Wilkie Collins' *No Name*. *Blackwood's Magazine* 94 (1863): 170.
Ollier, Edmund. 'Ghostly Pantomimes'. *Household Words: A Weekly Journal, Conducted by Charles Dickens* 7 (1853): 397–400.
'Faces'. *Household Words: A Weekly Journal, Conducted by Charles Dickens* 10 (1854): 97–100.
Olmstead, John C. *Victorian Painting: Essays and Reviews*. 2 vols. New York: Garland Publishing, 1980.
Outram, Dorinda. *Georges Cuvier: Vocation, Science and Authority in Post-Revolutionary France*. Manchester: Manchester University Press, 1984.
Paley, William. *Natural Theology, or Evidences of the Existence and Attributes of the Deity, Collected from the Appearances of Nature: together with: A Defence of the Considerations on the Propriety of Requiring a Subscription to Articles of Faith: Reasons for Contentment, Addressed to the Labouring Part of the British Public.* Vol. I of *The Works of William Paley*. London: N. Hailes, 1825.
Natural Theology, or, Evidences of the Existence and Attributes of the Deity, Collected from the Appearances of Nature. Ed. Charles Bell and Henry Brougham. London, 1836.
Paradis, J. and T. Postlewait, eds. *Victorian Science and Victorian Values: Literary Perspectives*. New York: New York Academy of Sciences, 1981.
Parris, Lesley, ed. *The Pre-Raphaelite Papers*. London: Tate Gallery, 1984.
Parsons, James. 'Human Physiognomy Explain'd: in the Crounian Lectures on Muscular Motion for the year MDCCXLVI'. *Philosophical Transactions of the Royal Society* 44 (1747): 1–82.
Pearce, Lynn. *Woman/ Image/ Text: Readings in Pre-Raphaelite Art and Literature*. Hemel Hempstead: Harvester Wheatsheaf, 1991.
Pearson, Karl. *The Life, Letters and Labours of Francis Galton*. 2 vols. Cambridge: Cambridge University Press, 1924.
Pick, Daniel. *Faces of Degeneration: A European Disorder, c. 1848–1918*. Cambridge: Cambridge University Press, 1989.
Piper, David. *The English Face*. London: Thames and Hudson, 1957.
Pointon, Marcia, ed. *Pre-Raphaelites Re-Viewed*. Manchester: Manchester University Press, 1989.
Porter, Roy. *The Making of Geology: Earth Science in Britain, 1660–1815*. Cambridge: Cambridge University Press, 1977.
'Making Faces: Physiognomy and Fashion in Eighteenth-Century England'. *Etudes Anglaises* 4 (1985): 385–9.
Potts, Alex. *Flesh and the Ideal: Winckelmann and the Origins of Art History*. New Haven: Yale University Press, 1994.
Price, F. 'Imagining Faces: The Later Eighteenth-Century Heroine and the Legible Universal Language of Physiognomy'. *British Journal of the Eighteenth Century* 6 (1983): 1–16.
Price, Rev. Thomas. *An Essay on the Physiognomy and Physiology of the Present Inhabitants of Britain; with reference to their origin, as Goths and Celts. Together*

with remarks upon the Physiognomical Characteristics of Ireland and some of the neighbouring continental nations. London: J. Rodwell, 1829.

Psomaides, Kathy A. 'Beauty's Body: Gender, Ideology and British Aestheticism'. *Victorian Studies* 36 (1992): 31–52.

Rabin, Lucy. *Ford Madox Brown and the Pre-Raphaelite History-Picture.* New York: Garland Publishing, 1978.

Ramsay, George. *Analysis and Theory of the Emotions, with dissertations on Beauty, Sublimity and the Ludicrous.* Edinburgh, 1848.

Redfield, James W. *Twelve Qualities of the Mind.* London, 1850.

Outlines of a New System of Physiognomy, illustrated by numerous engravings, indicating the Location of the Signs of the Different Mental Faculties. London: Webb, Millington, & Co., 1856.

Reynolds, Joshua. *Discourses on Art.* Ed. Robert R. Wark. New Haven: Yale University Press, 1975.

Richards, Evelleen. 'Darwin and the Descent of Woman'. *The Wider Domain of Evolutionary Thought.* Ed. David Oldroyd and Ian Langham. Dordrecht: Reidel, 1983. Pp. 57–111.

'Redrawing the Boundaries: Darwinian Science and Victorian Women Intellectuals'. *Victorian Science in Context.* Ed. Bernard Lightman. Chicago: University Of Chicago Press, 1997.

Richards, Robert J. 'The Influence of Sensationalist Tradition on Early Theories of the Evolution of Behaviour'. *Journal of the History of Ideas* 40 (1979): 85–105.

Ridley, Mark. *Evolution and Classification: The Reformation of Cladism.* London: Longman, 1986.

Ritvo, Harriet. 'New Presbyter or Old Priest?: Reconsidering Zoological Taxonomy in Britain, 1750–1840'. *History of the Human Sciences* 1 (1990): 259–76.

The Platypus and the Mermaid and Other Figments of the Classifying Imagination Cambridge, Mass.: Harvard University Press, 1997.

Rogerson, Brewster. 'The Art of Painting the Passions'. *Journal of the History of Ideas* 14 (1953): 68–94.

Rose, Andrea. *Pre-Raphaelite Portraits.* Yeovil: Oxford Illustrated Press, 1981.

Rossetti, Dante Gabriel. *The Poetical Works of D. G. Rossetti.* Ed. W. M. Rossetti. London: Ellis & Elvey, 1900.

Letters of Dante Gabriel Rossetti. Ed. Oswald Doughty and John Robert Wahl. 4 vols. Oxford: Clarendon Press, 1965–7.

Rossetti, William Michael. *Fine Arts, Chiefly Contemporary.* London: Macmillan, 1867.

The Pre-Raphaelite Journal: William Michael Rossetti's Diary of the Pre-Raphaelite Brotherhood, 1849–1853. Ed. William E. Fredeman. Oxford: Clarendon Press, 1975.

Rossetti, William Michael, ed. *Pre-Raphaelite Diaries and Letters.* London: Hurst and Blacket, 1900.

Ed. *Rossetti Papers 1862 to 1870*. New York: AMS Press, 1970.

Ruskin: Rossetti: PreRaphaelitism. Papers 1854 to 1862. New York: AMS Press, 1971.

Rothfield, Lawrence. *Vital Signs: Medical Realism in Nineteenth-Century Fiction*. Princeton: Princeton University Press, 1992.

Rupke, N. A. *The Great Chain of History: William Buckland and the English School of Geology, 1814–1849*. Oxford: Oxford University Press, 1983.

Ruse, Michael. *Darwinism Defended*. Reading, Mass.: Addison Wesley, 1982.

Sala, George. A. 'Getting up a Pantomime'. *Household Words: A Weekly Journal, Conducted by Charles Dickens*. 4 (1851): 289–96.

Sambrook, J. *Pre-Raphaelitism: A Collection of Critical Essays*. Chicago: University of Chicago Press, 1974.

Sawday, Jonathan. *The Body Emblazoned: Dissection and the Human Body in Renaissance Culture*. London: Routledge, 1995.

Schuster, John A., and Richard Yeo, eds. *The Politics and Rhetoric of Scientific Method. Historical Studies*. Dordrecht: D. Reidel, 1986.

Secord, James. 'Nature's Fancy: Charles Darwin and the Breeding of Pigeons'. *Isis* 72 (1985): 163–86.

Sekula, Alan. 'The Body and the Archive'. *October* 39 (1986): 3–64.

Seltzer, Mark. *Bodies and Machines*. New York: Routledge, 1992.

Shaw, A. *Narrative of the Discoveries of Sir Charles Bell in the Nervous System*. London: Longmans, 1839.

An Account of Sir Charles Bell's Discoveries in the Nervous System. London: Longman, 1860.

Reprint of Idea of A New Anatomy of the Brain, together with extracts of letters showing the origin and progress of his discoveries in the nervous system. London: Macmillan, 1868.

Shaw, Jennifer L. 'The Figure of Venus: Rhetoric of the Ideal and the Salon of 1863'. *Art History* 14 (1991): 540–70.

Shaw, John. 'An Account of Some Experiments on the Nerves; by M. Magendie: with Some Obervation by J. Shaw, Esq., Lecturer on Anatomy in Great Windmill Street'. *London Medical and Physical Journal* 48 (1822): 95–104.

'Remarks on M. Magendie's Late Experiments Upon the Nerves'. *London Medical and Physical Journal* 52 (1824).

Shookman, Ellis, ed. *The Faces of Physiognomy: Interdisciplinary Approaches to Johann Caspar Lavater*. Columbia, S.C.: Camden House, 1993.

Shortland, Michael. 'The Body in Question: Some Perceptions, Problems and Perspectives of the Body in Relation to Character c.1750–1850'. 2 vols. PhD diss., University of Leeds, 1984.

'Skin Deep: Barthes, Lavater, and the Legible Body'. *Economy and Society* 14 (1985): 273–312.

Siddons, Henry. *Practical Illustrations of Rhetorical Gesture and Action adopted to English Drama from a Work on the Subject by J. Engel, member of the Royal Academy of Berlin*. 2nd edn. London: Sherwood, Neely & Jones, 1822.

Smart, Alistair. 'Dramatic Gesture and Expression in the Age of Hogarth and Reynolds'. *Apollo* 82 (1965): 90–7.

Smith, Jonathan. *Fact and Feeling: Baconian Science and the Nineteenth-Century Literary Imagination.* Madison: The University of Wisconsin Press, 1995.

Smith, Lindsay. *Victorian Photography, Painting, and Poetry: The Enigma of Visibility in Ruskin, Morris, and the Pre-Raphaelites.* Cambridge: Cambridge University Press, 1995.

Smith, Roger. 'Physiological Psychology and the Philosophy of Nature in Mid-Nineteenth Century Britain'. PhD diss., University of Cambridge, 1970.

'The Background of Physiological Psychology in Natural Philosophy'. *History of Science* 11 (1973): 75–123.

Inhibition: History and Meaning in the Sciences of Mind and Brain. Berkeley: University of California Press, 1992.

The Fontana History of the Human Sciences. London: HarperCollins, 1997.

Spector, Benjamin. 'Sir Charles Bell and the Bridgewater Treatises'. *Bulletin for the History of Medicine* 12 (1942): 314–22.

Spencer, Herbert. 'The Haythorne Papers No. VIII. & No. IX. Personal Beauty'. *Leader* 5 (1854): 356–67 and 451–2.

The Principles of Psychology. London: Longman, Brohn, Green, and Longmans, 1855.

'Physical Training', *British Quarterly Review* (1858–9).

Stafford, Barbara Maria. 'From "Brilliant Ideas" to "Fitful Thoughts": Conjecturing the Unseen in Late Eighteenth-Century Art'. *Zeitschrift fur Kunstgeschichte* 48 (1985): 329–63.

'Peculiar Marks'. Lavater and the Countenance of Blemished Thought'. *Art Journal* 46 (1987): 185–92.

Body Criticism: Imaging the Unseen in Enlightenment Art and Medicine. Cambridge, Mass.: The MIT Press, 1991.

Stanton, Mary Olmsted. *An Encyclopedia of Face and Form Reading.* London, 1895.

Stein, Richard. *The Ritual of Interpretation: The Fine Arts as Literature in Ruskin, Rossetti, and Pater.* Cambridge, Mass.: Harvard University Press, 1975.

Stevens, Peter F. 'Species: Historical Perspectives'. *Keywords in Evolutionary Biology.* Ed. Evelyn Fox Keller and Elisabeth A. Lloyd. Cambridge, Mass.: Harvard University Press, 1992. 302–11.

Stevenson, Lionel. *Darwin Among the Poets.* New York: Russell & Russell, 1963.

Stolnitz, Jerome. '"Beauty": Some Stages in the History of an Idea'. *Journal of the History of Ideas* 22 (1962): 185–204.

Summers, David. *The Judgement of Sense: Renaissance Naturalism and the Rise of Aesthetics.* Cambridge: Cambridge University Press, 1987.

Sussman, Herbert L. 'The Language of Criticism and the Language of Art: The Responses of Victorian Periodicals to the Pre-Raphaelite Brotherhood'. *Victorian Periodicals Newsletter* 19 (1973): 15–29.

Facts into Figures: Typology in Carlyle, Ruskin, and the Pre-Raphaelite Brotherhood. Columbus: Ohio State University Press, 1975.

Szathmary, Arthur. 'Physiognomic Expression Again'. *Journal of Aesthetics and Art Criticism* 25 (1967): 307–12.

Taylor, Jenny Bourne. *In the Secret Theatre of the Home: Wilkie Collins, Psychology, and Sensation Narrative.* London: Routledge, 1989.

Taylor, Jenny Bourne and Sally Shuttleworth, eds. *Embodied Selves: An Anthology of Psychological Texts, 1830–1890.* Oxford: Clarendon Press, 1998.

Teich, Mikulas and Robert M. Young, eds. *Changing Perspectives in the History of Science.* London: Heinemann, 1973.

The Germ: Thoughts Towards Nature in Poetry, Literature and Art. London: Aylott & Jones, 1850. Repr. Oxford: The Ashmoleum Museum, 1992.

The Oxford and Cambridge Magazine for 1856 – Conducted by Members of the Two Universities. London: Bell & Daldy, 1856.

Thesing, William B. *The London Muse: Victorian Poetic Responses to the City.* Athens, Ga: University of Georgia Press, 1982.

Todd, R. *Physiology of the Nervous System.* London: Merchant Singer, 1847.

Todd, R., ed. *Cyclopedia of Anatomy and Phyisology.* 5 vols. London: Longman, 1835–59.

Topham, Jonathan R. 'Science and Popular Education in the 1830s: The Role of the Bridgewater Treatises'. *British Journal for the History of Science* 25 (1992): 397–430.

' "An Infinite Variety of Arguments": The *Bridgewater Treatises* and British Natural Theology in the 1830s'. PhD diss., University of Lancaster, 1993.

Tytler, Graeme. *Physiognomy in the European Novel: Faces and Fortunes.* Princeton: Princeton University Press, 1982.

Varley, John. *A Treatise on Zodiacal Physiognomy; illustrated by Engravings of Heads and Features; and accompanied by Tables of the Time of Rising of the Twelve Signs of the Zodiac; and containing also New and Astrological Explanations of some remarkable portions of Ancient Mythological History.* London: John Varley, 1828.

Walker, Alexander. *Physiognomy Founded on Physiology and Applied to Various Countries, Professions, and Individuals.* London: Smith, Elder & Co., 1834.

Beauty: Illustrated Chiefly by an Analysis and Classification of Beauty in Woman. Preceded by a Critical View of the General Hypotheses Respecting Beauty, by Hume, Hogarth, Burke, Knight, Alison, etc., and also followed by a similar view of the Hypotheses of Beauty in Sculpture and Painting, by Leonardo da Vinci, Winckelmann, Mengs, Bossi, etc. London: Henry G. Bohn, 1837.

Intermarriage; or, the mode in which, and the causes why, beauty, health, and intellect result from certain unions, and deformity, disease and insanity from others. London, 1838.

Woman, Physiologically Considered as to Mind, Morals, Marriage, Matrimonial Slavery, Infidelity and Divorce. London: A. H. Baily & Co., 1839.

The Book of Beauty: With Modes of Improving and Preserving It in Man and Woman. New York, 1851.

Walker, Alexander, Mrs. *Female Beauty, as Preserved and Improved by Regimen, Cleanliness and Dress; and especially by the Adaptation, Colour and Arrangement of Dress, as variously influencing the Forms, Complexion, & Expression of each individual, and rendering Cosmetic Impositions Unnecessary.* London: Thomas Hurst, 1837.

Walker, Donald. *Exercises for Ladies: Calculated to Preserve and Improve Beauty.* 2nd edn. London: Thomas Hurst, 1837.

Wallace, Alfred Russel. 'The Origin of Human Races and the Antiquity of Man deduced from the Theory of Natural Selection'. *Journal of the Anthropological Society of London* 2 (1864): clvii–clxxxvii.

'Mimicry and other Protective Resemblances among Animals'. *Westminster Review* 88 (1867): 1–20.

Contributions to the Theory of Natural Selection. London: Macmillan, 1870.

'[Review of] *The Descent of Man and Selection in relation to Sex.* by Charles Darwin'. *The Academy* 2 (1871): 177–83.

'[Review of] *The Expression of the Emotions.* by Charles Darwin'. *Quarterly Journal of Science* 3 (1873): 113–18.

Wallis, Roy, ed. *On the Margins of Science: The Social Construction of Rejected Knowledge.* Sociological Review Monographs, 27. University of Keele, 1979.

Walmsley, Edward. *Physiognomical Portraits.* 2 vols. London: J. Major, 1824.

Wechsler, Judith. *A Human Comedy: Physiognomy and Caricature in Nineteenth-Century Paris.* London: Thames and Hudson, 1982.

Wells, Samuel. *New Physiognomy; or, Signs of Character, as manifested through Temperament and External Forms, and especially in 'The Human Face Divine'.* London: L. N. Fowler & Co., 1866.

How to Read Character: A New Illustrated Handbook of Phrenology and Physiognomy for Students and Examiners. New York: S. R. Wells, 1873.

Whewell, William. *History of the Inductive Sciences.* 3 vols. London: J. W. Parker, 1837. 3rd edn. 3 vols. London: J. W. Parker & Sons, 1857.

Philosophy of the Inductive Sciences. 3 vols. London: J. W. Parker, 1840. 2nd edn. 2 vols. London: J. W. Parker & Sons, 1847.

Williams, Carolyn. 'The Changing Face of Change: Fe/Male In/Constancy'. *British Journal for Eighteenth Century Studies* 12 (1989): 13–28.

Winter, Alison. 'The Construction of Orthodoxies and Heterodoxies in the Early Victorian Life Sciences'. *Victorian Science in Context.* Ed. Bernard Lightman. Chicago: University of Chicago Press, 1997.

Mesmerised: Powers of Mind in Victorian Britain. Chicago: University of Chicago Press, 1998.

Woodring, Carl. *Nature into Art: Cultural Transformations in Nineteenth-Century Britain.* Cambridge, Mass.: Harvard University Press, 1989.

Woolnoth, Thomas. *Facts and Faces, or, the Mutual Connexion between Linear and Mental Portraiture morally considered.* London, 1835.

The Study of the Human Face. London, 1865.

Yeo, Richard. 'Science and Intellectual Authority in Mid-Nineteenth Century Britain: Robert Chambers and *Vestiges of the Natural History of Creation*'. *Victorian Studies* 28 (1984): 5–31.

'Scientific Method and the Rhetoric of Science in Britain, 1830–1917'. *The Politics and Rhetoric of Scientific Method. Historical Studies*. Ed. John A. Schuster and Richard R. Yeo. Dordrecht: D. Reidel, 1986. Pp. 259–97.

Defining Science: William Whewell, Natural Knowledge, and Public Debate in Early Victorian Britain. Cambridge: Cambridge University Press, 1993.

Young, Robert M. 'Malthus and the Evolutionists: The Common Context of Biological and Social Theory'. *Past and Present* 43 (1969): 109–45.

'The Role of Psychology in the Nineteenth-Century Evolutionary Debate'. *Historical Conceptions of Psychology*. Ed. M. Henle, J. Jaynes and J. J. Sullivan. New York: Springer, 1973.

Darwin's Metaphor: Nature's Place in Victorian Culture. Cambridge: Cambridge University Press, 1985.

Mind, Brain and Adaptation in the Nineteenth-Century: Cerebral Localization and the Biological Context from Gall to Ferrier. Oxford: Oxford University Press, 1990.

Youngson, A. J. *The Scientific Revolution in Victorian Medicine*. New York: Holmes, 1979.

Index

anatomy 46–7, 61–6, 80–1, 84, 90–1, 116
Aristotle 4, 17, 53

Bain, Alexander 5–6, 114–15, 121, 139,
145–6, 152, 156, 169–70, 178–9; on
associationism 123–6; *Senses & Intellect* 3–4
Barasch, Mosche 91–2, 95–6, 103
Barrell, John 208 n. 36
beauty: and healthy physiques 13–14, 110–14,
185–6; of expression (facial) 67–9, 128–30,
135–41; and morality 110–12, 120–3,
139–40
Beer, Gillian 49–50, 60
Bell, Charles 4, 6–7, 13, 36, 43, 91–2, 108,
115, 152, 156–7, 165, 178–9; on natural
theology 44–5, 52–4, 58–60; on sensation
and creation 47–8, 55–8; and Johann
Caspar Lavater 46–7; on the nerves 55–8;
on art and anatomy 60–6; on grief 74–5;
on blushing 76–7; *System of Dissections* 46;
Anatomy of Expression 47, 61–79, 80–2,
143–5, 198 n. 11; *The Hand* 47, 52, 58–61;
'Idea of Brain' 47, 55–8
Bewick, William 65–6
Blumenbach, J. F. 71
blushing 76–7, 167–78
Boccaccio, Giovanni 85
Bridgewater treatises 44–5
British Association 182
Bronkhurst, Judith 94
Brougham, Henry 46
Brown, Ford Madox 79, 101, 106–9; on
pantomimic art 99–100; *Last of England*
106–7
Browne, Janet 144
Burgess, Thomas 152, 174–6
Butler, Joseph (Bishop) 49–50

Camper, Pieter 71–3, 152
Carpenter, William B. 5, 145–6, 156, 164–5,
178–9

character: type of 2–3, 38–40, 183–4;
investigation of 14, 37, 110–15, 180–8; and
appearance 110–15, 126–30
Clarke, Edwin 26, 165
Clarke, W. T. 139–41
Codel, Julie F. 90
Collins, Wilkie 115, 126–41; *Woman in White*
126–30; *Basil* 126, 132–6; *No Name* 126,
137–9
Collinson, James 81
consciousness 26, 114, 123–5, 132–7, 173–8;
and contrivance 43
Cooper, Robyn 115
Cowling, Mary 5–6, 107
Cross, John 15, 17–19
Cuvier, Georges 5, 56, 71–2; on comparative
anatomy 53–4; and Paley 54–5
Cvetkovich, Ann 131–2

Darden, Lindley 37
Da Vinci, Leonardo 116
Darwin, Charles R. 4–6, 11–12, 14, 36, 76,
79; on Charles Bell 143–6, 149–50, 152,
213 n. 14; scepticism for physiognomy
145–6; on principles of expression 156–61,
215 n. 32; on grief 159–61, 215 n. 31; on
blushing 167–78; *Expression of the Emotions*
142–6, 149–51, 159–61, 174–9; *Origin of
Species* 142, 162; *Descent of Man* 142, 168, 171;
Voyage of Beagle 168–73
Darwin, Erasmus 30–1, 156
De Piles, Roger 20
Descartes, René 19–27, 102, 195 n. 18;
Discourse on Method 21; *Passions of Soul* 21
design: concept of 13, 50–2, 174; influence on
Bell 45, 52–3, 58–61
Desmond, Adrian 5, 54–5, 190–1 n. 12
Dickens, Charles, *Household Words* 103–4
Diderot, Denis 100
dualism 20–7; and monism 31, 57, 166
Duchenne, G. B. 152–6

240

CAMBRIDGE STUDIES IN NINETEENTH-CENTURY
LITERATURE AND CULTURE

General editor
Gillian Beer, *University of Cambridge*

Titles published